10/13

BIG GUNS
BRAVE MEN

BIG GUNS

MOBILE ARTILLERY OBSERVERS

BRAVE MEN

AND THE BATTLE FOR OKINAWA

RODNEY EARL WALTON

NAVAL INSTITUTE PRESS
ANNAPOLIS, MARYLAND

Naval Institute Press
291 Wood Road
Annapolis, MD 21402

Library of Congress Cataloging-in-Publication Data

Walton, Rodney Earl.
Big guns, brave men : mobile artillery observers and the battle for Okinawa / Rodney Earl Walton.
pages cm
Includes bibliographical references and index.
ISBN 978-1-61251-130-6 (hardcover : alk. paper) —
ISBN 978-1-61251-131-3 (e-book) 1. World War, 1939–1945—Campaigns—Japan—Okinawa Island. 2. World War, 1939–1945—Artillery operations, American. 3. United States. Army—Artillery—History—20th century. 4. Artillerymen—United States—History—20th century. I. Title.
D767.99.O45W35 2013
940.54'252294—dc23
2012048194

This paper meets the requirements of ANSI/NISO z39.48-1992 (Permanence of Paper).

Printed in the United States of America.

21 20 19 18 17 16 15 14 13 9 8 7 6 5 4 3 2 1

First printing

Book design and composition: David Alcorn, Alcorn Publication Design

To the members of the U.S. Army artillery forward observation teams and artillery liaison teams who made the ultimate sacrifice for their nation during the battle for Okinawa, April–June 1945

CONTENTS

MAPS AND CHARTS

MAPS

CHARTS

ACKNOWLEDGMENTS

Like many historical accounts, this book could not have been written without the assistance of numerous people. Dr. Boyd L. Dastrup, command historian at the U.S. Army Fires Center of Excellence at Fort Sill (Oklahoma), was particularly gracious in providing unpublished manuscripts from the Morris Swett Library at Fort Sill. Dr. Dastrup, a leading authority on the history of artillery, verbally supplemented two of his published accounts with background information on the development of the American artillery forward observer during the period between World War I and World War II.

Because it relies largely on oral history, this work could not have been created without the willingness of several World War II veterans to discuss their recollections with me. A list of the veterans can be found at the end of this work. My thanks go out to all of them.

In light of the number of casualties on Okinawa, I felt particularly fortunate to have located a few important sources (including Sheahan, Moynihan, Bollinger, Knutson, Thompson, and Walton) who participated in the entire battle from the beginning to the end. Of these, only Walton received even a minor wound, and he was never out of action. Of course, sources who served only a portion of the campaign also made significant contributions toward my understanding of the battle.

One source important throughout this account is a series of letters written by artillery forward observer Ray D. Walton Jr. (Lieutenant Walton), my father. Only one of the letters was written while the battle of Okinawa was under way, and the information it contains was, for security and censorship reasons, relatively unimportant. Walton, however, prepared another series of letters to members of his family beginning early in 1995 (the fiftieth anniversary of the battle) that recount events before, during, and after the battle. Walton's letters offer a clear view of the human aspects of a forward observer's daily life in the field during the Okinawa campaign.

I interviewed Walton on videotape in 1993 following his first return visit to Okinawa. He was interviewed again in June 1995 during his second return visit as part of the commemoration ceremonies. A portion of this second interview, conducted by Associate Professor A. P. Jenkins of the University of the Ryukyus, was published in a predominantly Japanese-language booklet containing memories of the fifteen years during which Japan was engaged in war with China and other nations during the 1930s and 1940s.

Walton's memories of Okinawa are unfortunately incomplete. Other than the April 1, 1945, landings and the April 10, 1945, assault on Kakazu Ridge, he could recollect little of his observation team's role in the seizure of key objectives of the 96th Infantry Division such as Hacksaw (or Sawtooth) Ridge, Conical Hill, the Big Apple, and the Medeera pocket. He was unable to supply dates and locations for the photographs from his collection that are included in this account, although the dryness of the landscape eliminates the period when there were heavy rains (late May–early June 1945).

Two important members of Walton's forward observation team, Fred Goebel and George Arnold, were believed to be deceased and therefore could not be interviewed, but I did have access to two of Goebel's letters from June 1945. Goebel was an enlisted man who was the second in command of Walton's field observation team for most of the battle. The letters say little about the battle but are significant in that they demonstrate the casualties being taken by the artillerymen of B Battery. The letters also reveal something about the relationship between officers and men in the forward observation teams.

I conducted a lengthy interview with Charles Sheahan, an artillery officer who was from the same artillery battalion as Walton and provided artillery support to the same 381st Infantry Regiment. I also interviewed Donald Burrill, one of Sheahan's subordinate observers, concerning his heroic conduct early in the campaign. Charles P. Moynihan's interview provides the point of view of an enlisted artilleryman. In addition to Bollinger, I interviewed two other infantrymen from the 2nd Battalion, both from the enlisted ranks.

Two visits to the battlefields of Okinawa, accompanied on both occasions by Walton, contributed details about the battlefield sites. During the first visit, in 1993, we were privileged to have as our guide Ms. Setsuko Inafuku, a native Okinawan who had survived the battle as an infant. Because there is still a large contingent of Marines on Okinawa, this guide normally focused her tours on the areas where the Marines fought, but she was kind enough to take us to some of the areas where the Army fought as well.

A second trip, in 1995, on the fiftieth anniversary of the battle, focused on areas where the 96th Division had fought. This included a visit to the nearby island of Ie Shima, where the 77th Division fought and where the famous newspaper correspondent Ernie Pyle died. The tour was sponsored by Valor Tours and guided by U.S. servicemen stationed on the island of Okinawa. I particularly want to thank Donald Dencker, a 96th Division veteran of the battle and later the author of an infantryman's memoir of the battle.

Several members of the 1995 tour group were veterans of the 96th Division. During this trip Curt Sprecher, an infantryman from the battalion supported by Lt. Walton, discussed his recollections of the battle on the very site where he had been when an artillery observation officer was killed nearby.

Also accompanying the tour group were William C. Buckner and Mary Buckner Brubaker, the son and daughter of Lt. Gen. Simon Bolivar Buckner Jr. who kindly made available to tour members the general's letters written in close proximity to and during the Okinawa campaign. The 96th Division Association (especially treasurer Robert Schmidt and historian Donald Dencker) was helpful in directing me to the appropriate veterans to interview. Although they may not have agreed with all of my conclusions, Dencker, retired Marine Corps colonel Joseph Alexander, and attorney Alan Christenfeld provided helpful assistance by commenting on a draft of my dissertation. My sister-in-law Peggy Walton helped proofread one draft. Although Dr. Eric Leed retired from the Florida International University History Department before I had begun the Ph.D. program and thus did not serve on my dissertation committee, he nonetheless introduced me to oral history and some of its techniques by allowing me to assist on one of his projects during the 1990s.

This book has its origins in my graduate work at Florida International University. Scholars seeking more detailed documentation of this account may wish to refer to my dissertation, which is listed in the bibliography. Dr. Darden Pyron, the original founding member of the History Department, undertook the onerous task of directing my research project and serving as chair of my dissertation committee. He had faith in the academic legitimacy of this project long before I did. More than a dozen years ago (and well before I became a Ph.D. student in history), Dr. Pyron encouraged me to consider expanding my research on Okinawa. Each dissertation committee member contributed as well. Dr. Gwyn Davies encouraged me to examine the development of the forward observer within the artillery branch. Dr. Kenneth Lipartito encouraged me to enrich the social history of the Okinawa campaign by examining the daily life experiences of the observers. Dr. Ralph Clem, a distinguished military veteran, encouraged me to examine the impact of friendly fire in the crucible of the Okinawa campaign. I am indebted to the entire committee. Any errors, of course, are mine alone.

BIG GUNS
BRAVE MEN

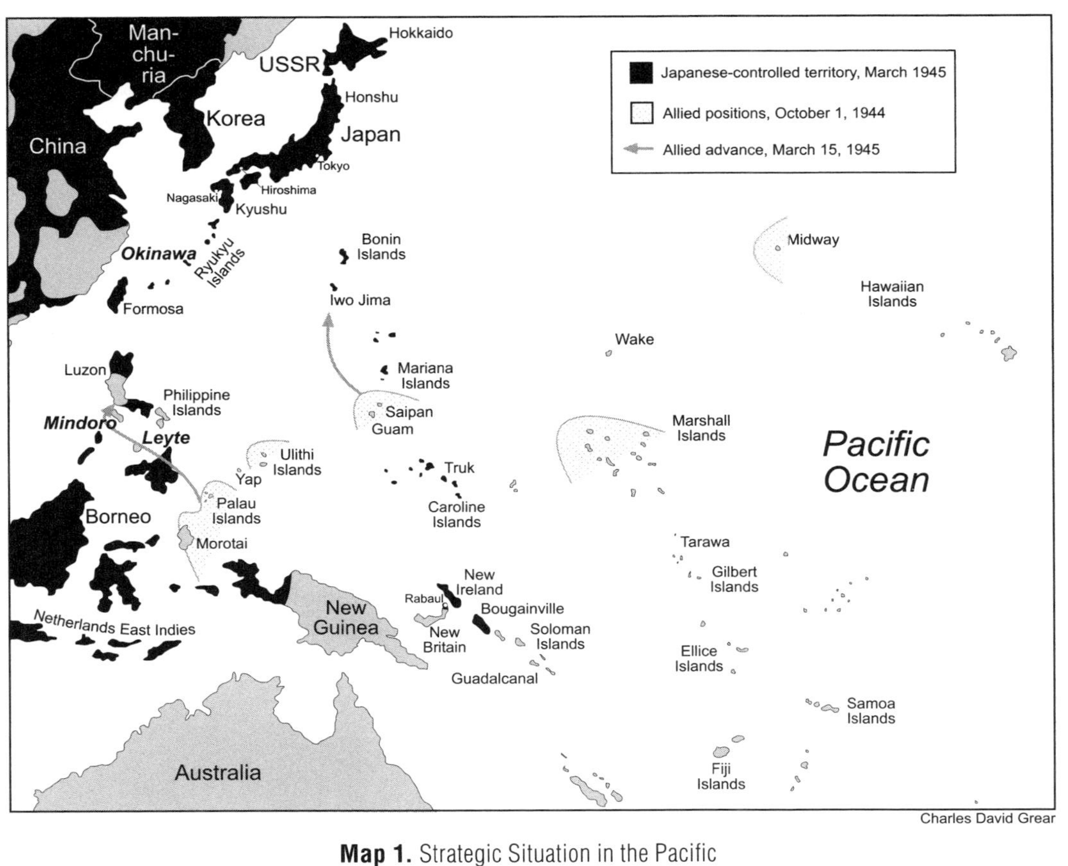

Map 1. Strategic Situation in the Pacific

Adaptation by Charles Grear based on map I in Roy E. Appleman, James M. Burns, Russell A. Gugeler, and John Stevens, *Okinawa: The Last Battle* (Washington, D.C.: Center of Military History, 1948, 1991).

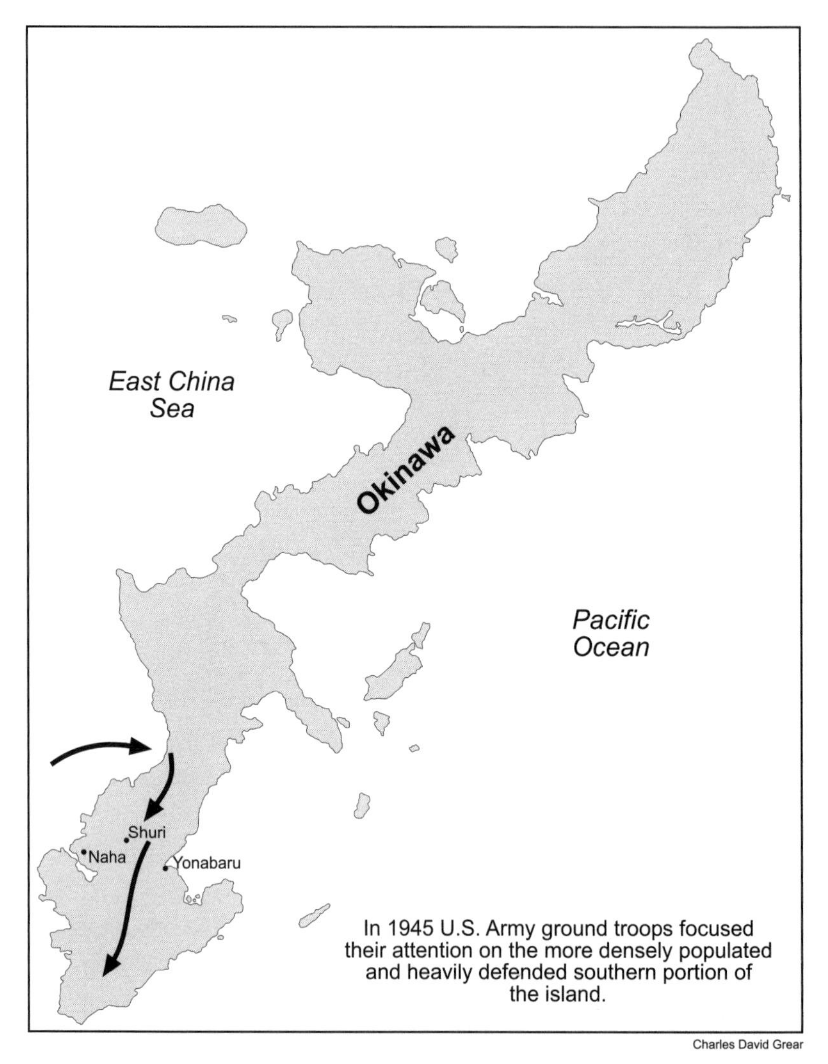

Map 2. Okinawa

Adaptation by Charles Grear based on map 14 in Orlando R. Davidson, J. Carl Willems, and Joseph A. Kahl, *The Deadeyes: The Story of the 96th Infantry Division* (Nashville: Battery Press, 1947, 1981).

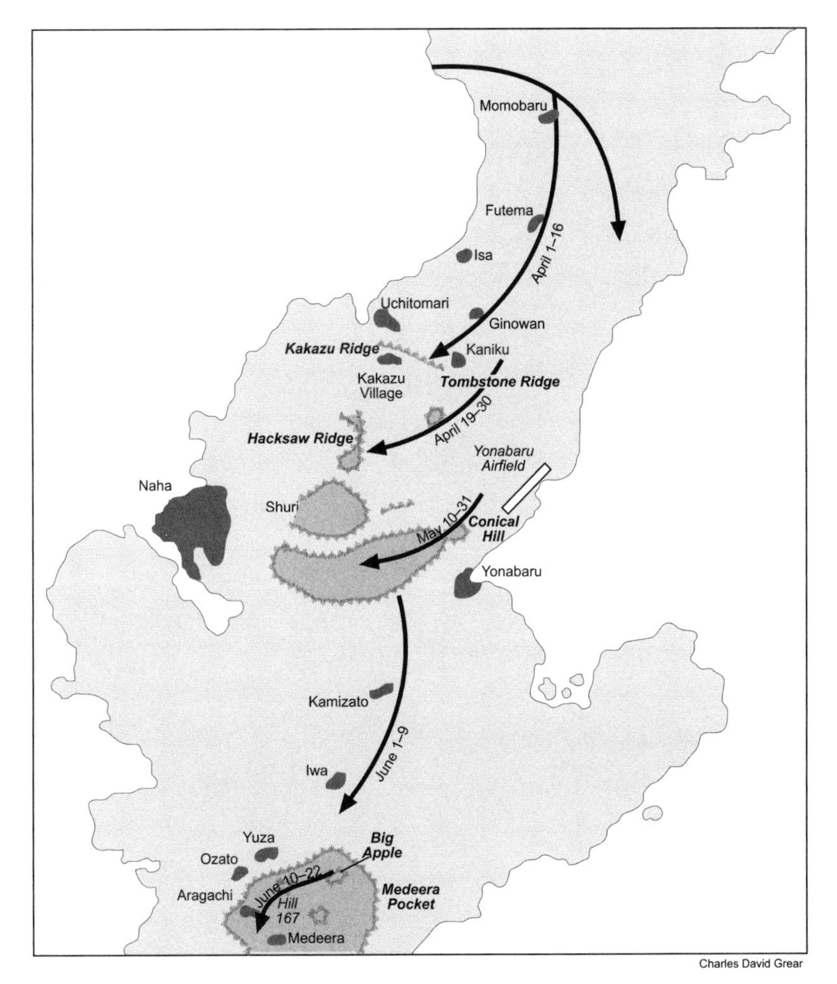

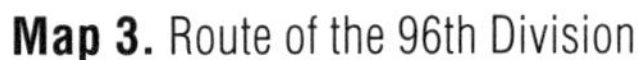

Map 3. Route of the 96th Division

Adaptation by Charles Grear based on map 15 in Orlando R. Davidson, J. Carl Willems, and Joseph A. Kahl, *The Deadeyes: The Story of the 96th Infantry Division* (Nashville: Battery Press, 1947, 1981).

INTRODUCTION

World War II (1939–45) was the largest and deadliest conflict in history. The Pacific theater alone was a massive undertaking. The forces that fought there—Japanese, Americans, and their respective allies—mobilized millions of people who fought over a vast area one and a half times the size of the European theater for almost four years. Following the devastating attack at Pearl Harbor in 1941, American forces fought their way across the Central and Southern Pacific, island by island, in the face of staunch Japanese resistance. When the great conflict was finished, twenty million people had perished in the China-Pacific theater.[1]

Of the many Pacific battles, Okinawa stands out as both the final major American land battle of World War II and the greatest air-sea battle in history. Japanese resistance lasted more than eighty days, between April and June 1945. The outcome of the Pacific war was at stake; Okinawa would provide American forces with the naval port and air bases necessary for the anticipated invasion of Japan. Okinawa has also been called "the bloodiest land battle of the Pacific war"; 12,281 Americans died there, and the total U.S. casualties, including wounded and nonbattle casualties, numbered more than 70,000.[2] The official U.S. Army history of the battle estimated 110,071 Japanese deaths with only 7,401 Japanese prisoners captured. An estimated 150,000 Okinawan civilians were killed as well.[3]

A half century after the battle, an American strategic analyst characterized the Japanese commander, Lt. Gen. Mitsuru Ushijima, and his forces on Okinawa as "the best the Japanese had."[4] Col. Hiromichi Yahara, one of the highest-ranking Japanese officers to survive the battle, described it as "the worst fighting of the Pacific war, its sustained intensity surpassing even the brutal combat of Tarawa, Peleliu, and Iwo Jima."[5] Okinawa was the only battle of the Pacific war in which the commanding generals on both sides lost their lives.

Field artillery played a critical role in the battle. Although the Japanese army had traditionally emphasized light infantry tactics, Okinawa signaled a shift toward artillery.[6] The Japanese dispatched one of their most respected artillery experts, Lt. Gen. Wada Kosuke, to command the island's big guns.[7] They also concentrated more artillery pieces on Okinawa than anywhere else in the Pacific.[8] Lt. Gen. Simon Bolivar Buckner Jr., the American commander on Okinawa and the highest-ranking U.S. officer to die in combat during World War II, was killed by Japanese artillery fire.[9]

In the event, U.S. artillery matched the Japanese in intensity and surpassed it in effectiveness.[10] It was no accident that American artillery proved superior. The round-the-clock presence of artillery observers provided American infantry with much faster artillery support than the six-hour time lag reported for the Japanese army.[11] The U.S. Army's use of forward artillery observers was at least in part an effort to avoid the artillery "friendly fire" problems that had occurred in some armies during World War I. This book examines one source of American artillery superiority on Okinawa, in particular, frontline artillery support. More specifically, it analyzes the role of U.S. Army forward artillery observers, a hitherto understudied part of the battle and of the war itself.

Artillery observers were for the most part mid-level, middle-class officers without a great deal of combat experience but intelligent enough to make the calculations necessary to provide accurate artillery support. They wielded a deadly weapon that could inflict huge damage on friendly as well as enemy forces. Like the infantrymen they worked alongside, they faced constant danger. Complex relationships form when relatively inexperienced mid-ranking men from a different combat arm are imbedded within a veteran infantry unit (with little tolerance for error) in a combat situation.

This book is intended to fulfill six goals. The first goal is to supplement *Okinawa: The Last Battle*, by Roy Appleman, James Burns, Russell Gugeler, and John Stevens, which despite the passage of sixty years remains the Bible of the campaign.[12] An early reviewer noted that the book represented "the ultimate development of . . . methods employed by the War Department to obtain historical coverage of operations."[13] Although Appleman's work bears the taint of being an "official" history, it represents the apex of that type of account. Another prominent reviewer noted that "Okinawa was not only the last and greatest land battle in the Pacific; it was also the first in which trained historical personnel were attached to divisions, corps, and army."[14] Maj. Roy Appleman served on the scene as the historian for XXIV Corps, the headquarters for all U.S. Army maneuver units on the island. Capt. Charles Sheahan called Appleman's coverage of the 381st Infantry Regiment—the unit Sheahan had supported—"very complete."[15] Appleman's extensive maps alone are worth the price of the book.

Appleman's text has impressive strengths—particularly in its coverage of infantry and armor operations. It emphasizes tank-infantry teams as central to the American victory, citing the importance of the "flame and demolition that destroyed the Japanese in their strongholds." Bob Green, an armor lieutenant in the battle of Okinawa, wrote in 2004 that "the use of these diabolical machines [flame-throwing tanks] contributed more to the defeat of the Japanese on Okinawa than any other American weapon." General Buckner called this method of combining liquid flame and explosives the "blowtorch and corkscrew" method.[16] Most of the texts written after Appleman's, including in the twenty-first century, likewise reflect the viewpoint of the infantrymen who fought on Okinawa and emphasize the armor and infantry aspects of the battle.[17]

Despite its longevity as the classic account of the battle, Appleman's book has its weaknesses. One is that it downplays the role of artillery relative to that of infantry. Chester Starr, a prominent historian of ancient Europe and a historian of the U.S. Army's World War II Italian campaign, criticized the book for giving the impression that "artillery preparations were of dubious value; only when an observer could actually see a specific target were even 16-inch battleship shells of use."[18] James A. Field Jr. similarly questioned Appleman's treatment of artillery, noting Appleman's failure to indicate how commanders requested artillery fire and how quickly the artillery responded.[19] Another reviewer also complained about Appleman's failure to highlight the artillery's contributions to the battle.[20]

Such criticisms are valid. General Buckner followed orthodox U.S. Army doctrine, which relied on the use of heavy artillery to eliminate enemy resistance.[21] On April 22, 1945, for example, Buckner wrote to his wife: "The artillery is beginning to roar down at the front and parachute flares are lighting up the whole sky. This probably means that the Japs are putting on a night counter-attack. I hope so since this will bring them out of their caves and we can use our artillery on them with good effect."[22] The battle of Okinawa, according to the authors of the 96th Infantry Division history, thus "produced the greatest artillery duels in the war against the Japanese. Newspapermen placed it on par with the great artillery battles in the European theater."[23] Indeed, Buckner critics such as Gen. Joseph "Vinegar Joe" Stilwell objected to his "heavy emphasis on artillery."[24]

The second goal, in addition to giving artillery its rightful place in the battle for Okinawa, is to illuminate the Okinawa campaign from the point of view of the men who provided frontline artillery support to U.S. Army infantry units. Historian Orlando Davidson called these men "among the least recognized heroes of the war," and K. P. Jones, a forward observer in the European theater, agreed that "little has been said or written about their exploits." Yet, two academic historians writing a survey of World War II described the forward observation team as "probably the single most effective killer in ground combat," and Boyd Dastrup notes that "the appearance of new fire control techniques and portable field radios provided the most striking difference between World War II and World War I field artillery capabilities."[25]

Artillery observers operated in small groups called forward observer teams and liaison teams that were stationed near the line of battle itself. The mobile forward observer not only served as the eyes of the artillery, he also helped other service branches as well. Even in places like Okinawa, where the extensive Japanese underground fortifications lessened the effect of artillery, a combined arms approach was an essential ingredient to victory. As the only artilleryman who could see what the maneuver commander was seeing, the forward observer thus tied together the various combat arms.[26] The forward observer had to understand not only how the artillery functioned but also how the infantry (or armor) operated.[27] Without his coordinating efforts, forward progress on the battlefield was difficult indeed. Earlier books on the subject of the American forward observer treat the European–North African theater exclusively.[28] This book completes the picture.[29]

Third, this work explores the problem of friendly fire. When this project was conceived in the 1990s, I had no plans to cover this topic. The subject of "short rounds," however, developed into a critical part of the book. The issue was of obvious importance because the artillerymen themselves (as well as some of the infantrymen) consistently volunteered information on this point, citing friendly fire as a constant source of pressure on the observers. One 96th Division observer, a veteran of Leyte and Okinawa, recalled that it "was nerve-wracking fearing to fire on your own infantry. Thank God it never happened to me."[30] Furthermore, friendly fire was prominently discussed in the daily reports prepared by military units while the action was still raging in 1945; and yet the existing Okinawa literature largely ignores it.[31]

The fourth goal of this book is to correct a historical imbalance. Despite its magnitude, far less attention has been paid to the Okinawa campaign than to Iwo Jima, Guadalcanal, and Tarawa.[32] Even at the time it was taking place, the Okinawa campaign was overshadowed by such earth-shaking events as the death of President Franklin Roosevelt and the surrender of Germany. Shortly after the campaign ended, the Potsdam Conference, the atomic bomb, and the surrender of Japan immediately absorbed the public's attention. For too many people, the Pacific war consists primarily of Pearl Harbor, Iwo Jima, and the atomic bombs.

Fifth, this book addresses another kind of imbalance present even in the existing literature on Okinawa. Much of what has been written about the campaign centers on the activities of the U.S. Marine Corps.[33] Indeed, the classic account of a World War II infantryman is E. B. Sledge's *With the Old Breed at Peleliu and Okinawa* (1981). The naval aspects of the battle, which were substantial, have also been recounted.[34] Even the Japanese viewpoint has been published, despite the destruction of 92 percent of the Japanese army.[35] The land campaign on Okinawa, however, was primarily an Army battle and was commanded by an Army general.

No single book, of course, can cover the experiences of all the U.S. Army artillerymen on Okinawa. This work focuses on fire support rendered to one of the 96th Infantry Division's three infantry regiments, the 381st Infantry Regiment. An infantry regiment comprised roughly three thousand men. The unit report for April 1, 1945—the first day of the Okinawa campaign—shows the 381st to have been authorized 153 officers and 3,025 enlisted men, although it was short 7 officers and 340 enlisted men at that point. The regiment consisted of three battalions of infantry, the 1st, 2nd, and 3rd, plus a headquarters company, an antitank company, a cannon company, a service company, and a medical detachment.[36] The primary artillery fire support for the 381st came from the 361st Field Artillery Battalion, one of four field artillery battalions belonging to the 96th Division. These four battalions fired 302,852 rounds during the Okinawa campaign. The forward observers and liaison officers of the 361st often had available to them artillery fire from other sources as well, including adjacent units (Army, Marine Corps, and Navy), Army corps–level artillery, naval gunfire, and tactical air support.[37] This book is not intended to be a technical analysis of artillery support in the Okinawa campaign; instead it focuses on the experience of battle and tries to reconstruct the ambience and the chronology of frontline artillery combat on Okinawa.[38]

Sixth, this book examines the human experience of combat for the men who controlled the awesome firepower of World War II artillery.[39] What were their impressions, hopes, and fears? What were their successes and their failures? What concerns still haunted them when they recorded their memories more than half a century after the campaign ended? What was life like in the muddy foxholes and canvas tents of the Okinawa campaign? What were their relations with other members of the forward observation and liaison teams? What were their relations with the infantry units they supported—in particular with the commanders of those formations, who often set the agenda for the artillery support?

I will address these issues chronologically, beginning with the events leading up to the invasion of Okinawa and ending with the surrender of Japan in September 1945 as combat troops from Okinawa were being redeployed. The most detailed coverage is for the period April 1, 1945 (the invasion of Okinawa), to June 22, 1945 (the end of organized Japanese resistance on Okinawa).

Oral history is central to this account.[40] The use of oral interviews builds on a methodology established at least as early as the Greek military historians of the fifth century BC. The Greek historian Thucydides, who lived from about 460 BC to about 400 BC, used oral history to collect information for his famous account of the Peloponnesian War.[41] Only in the mid-twentieth century, however, did oral history come into frequent use in a modern military context, in conjunction with the invention of the portable tape recorder and the ready availability of educated personnel to act as interviewers. The type of modern industrial warfare fought in World War II resulted in massive casualties and had widespread societal impacts. Millions of ordinary individuals were conscripted or volunteered to fight, and each soldier, sailor, and air personnel had a unique experience. Modern techniques of oral history have made it possible to collect and archive their memories.

Interviews with veterans can yield fragments of military history long after a battle is over.[42] Military historians are well aware that memories fade and become less reliable over time, of course. Historian Stephen Everett noted that "interviews conducted months or years after the fact cannot replicate the freshness, accuracy, and details of those conducted within a few days of the actual events." Everett also recognized that "as time increases between an experience and its recounting, individuals tend to condense the sequence of events and omit critical actions and judgments."[43]

Battlefield memories may be more reliable than most in that regard. They are often burned into the veteran's mind, perhaps because of the adrenalin flow that battle generates.[44] Some veterans can recall details of a battle that took place a half century ago better than events of last week.[45] Veterans themselves have noted the unusual longevity of wartime memories. "War is a tragic—exhilarating experience," Charles Sheahan explained to me. "The memory of [it] sticks like glue in the back roads of our minds. The years roll along decade after decade in our civilian pursuits. However, time will never erase the times when one's body and soul were on the edge of the abyss."[46]

The veterans' own commitment to the oral history project gives an aura of legitimacy to their accounts. In order to accurately recount their experiences, some elderly artillerymen had to make enormous efforts to overcome the physical effects of aging. Forward observer Oliver J. Thompson reported of his efforts sixty-four years after the battle, "My eyes tear up and fingers ache in writing too much."[47]

A unique preexisting personal archive of interviews, primarily from the 1990s, forms the core of this book, but these piecemeal accounts had to be checked for accuracy and placed within the larger context of the entire battle. I checked the verbal accounts against a review of unit reports obtained from the U.S. National Archives and Records Administration and supplemented them with secondary sources prepared by historians. The Okinawa veterans' oral accounts given long after the battle are understandably incomplete, but the coverage by the military historians who accompanied the troops is excellent. In particular, Appleman's account provides a detailed overview of the strategy of the campaign. The book-length 96th Division history also provides important background information.[48] The records of the 361st Field Artillery Battalion and the 381st Infantry Regiment for the period April 1 to June 30, 1945, obtained from the National Archives, provided additional information. The records of the 361st Field Artillery Battalion were altogether inadequate. A skimpy "Battalion History" of that unit for the Okinawa campaign does little more than list casualties and decorations.[49] The National Archives documentation on the 381st Infantry Regiment is much better. The history of that unit, which naturally focuses on infantry action rather than artillery, consists of several pages and is very well written. The staff intelligence/operations journal and the daily unit reports for the 381st sometimes refer to artillery successes and failures.

The oral histories and the written record sometimes conflict, but there is frequently an acceptable explanation for the discrepancy. The oral account from the most highly decorated forward observer I interviewed, for example, Donald Burrill, does not agree with the record in several details. His description of an April 1945 event for which he was awarded a Silver Star generally coincided with the written recommendation and certificate for that award, but both the date and the place named in his oral account differ from the written record. Burrill insisted that the event took place on April 19, 1945, whereas the written record says April 28, 1945. Burrill said the incident occurred on Tombstone Ridge, whereas the record gives a location roughly two miles farther south at Sawtooth (Hacksaw) Ridge.

Burrill was aware of the discrepancy but considered it a minor error. He was seriously wounded during the incident and was immediately evacuated from Okinawa to Saipan. Thus, he was not present to verify the accuracy of the recommendation form when it was prepared on May 26, 1945—a month after the incident. He did not rejoin his unit until after Okinawa had been declared secure. The award ceremony was held soon after he was reunited with his unit. The veteran was on friendly terms with the infantry company commander who had written the recommendation, but he likely would not have felt comfortable contradicting the officer who had recommended him for the award.

In this instance I elected to use the date and place stated by the veteran while nonetheless noting the discrepancy with the written record. The 96th Division history, published two years after the war, confirms that Burrill received the Silver Star, and I was far more concerned about the experiences of the forward observers during the battle than the exact date and place of any specific event. Burrill gave no history of readjustment problems following the war and had no incentive to falsify the date and place. Indeed, because a Japanese bullet had inflicted a painful wound to his pelvis, the date of the incident was more likely seared into his mind than that of anyone else. One aspect of the recommendation form submitted by the company commander also supports Burrill's version. He located the incident near the town of Kaniku, which is much closer to Tombstone Ridge (the location given by Burrill) than it is to Sawtooth Ridge (the location listed in the documents).

Memory gaps remain a legitimate concern for oral interviewers for good reason. Veterans' memories of events that occurred earlier in their wartime experience may be more reliable than memories of events that took place later. For example, in interviewing 96th Division veterans of the 1945 Okinawa campaign, I found that those who had participated in the whole campaign generally had quite good memories of the first major battle (Kakazu Ridge), fair recall of the second major battle (Hacksaw Ridge, also known as Sawtooth Ridge), and poor memories of later battles (Conical Peak, the Big Apple, the Medeera pocket), even though, from a historian's perspective, the later battles brought the campaign to a dramatic conclusion.[50]

In the case of one veteran, two trips to Okinawa, interviews conducted on the sites of various battles, and listening to the accounts of other veterans of the campaign failed to resolve memory gaps. This pattern occurred even among artillery forward observers of the officer class, who consistently had access to maps and troop location information in order to perform their jobs. Such memory deficits were a bit surprising because the campaign lasted less than three months. One enlisted man, who later graduated from Berkeley, explained it to me this way:

> Now I want to tell you one thing from this point [Conical Peak] on. For some unknown reason I can't remember very much about anything from now [mid-May] on till I got relieved [mid-June]. I don't know why. . . . You'll find out I'll become kind of vague. . . . I never could do it. It just seems like the mind just closed down. I remember some things but I can't remember just—I get kind of lost. . . .You see what actually happens is for some unknown reason your mind starts cutting all this stuff out. And I think we reached a point that we couldn't handle it any more. And then our subconscious would go—you know just click it out.[51]

The problem of memory gaps is not particularly harmful as long as the veteran is candid about his memory problem (as most were). Replacement troops who arrived during the course of the campaign may have better recollections of the middle and later battles in the campaign because their minds had not yet begun to tune out the details of the battle.

Interviews conducted long after the battle also carry the risk that the veteran's recollection has been colored and influenced by media (movies, television, documentaries, books, and the stories of other veterans). Alistair Thomson claimed to have observed this phenomenon during his oral history interviews with Australian World War I veterans, noting that "in some interviews I felt like I was listening to the script of the film *Gallipoli.*"[52] I did not personally encounter this phenomenon, but that may be because the Okinawa campaign had not been the subject of a major movie, miniseries, or major documentary series at the time of the interviews. Everett recommended testing the accuracy of the veteran's memory by asking questions to which the interviewer already knows the answer. "An interviewee's response to these queries may shed new light on an issue," he noted; "if not, their answers may serve as yardsticks to judge the accuracy of other information provided by the interviewee."[53] Another method is to press for additional detail in an effort to find out if the veteran actually experienced an event or just "heard about it," a fairly frequent occurrence. In many cases, the veteran will have told his story many times, and what the interviewer hears is a well-rehearsed and polished version of it. In such instances, the interviewer must use detailed questions to separate the hearsay from the actual experience, if such is possible.

My experience interviewing veterans of World War II suggests that outside influence and overpolished presentations are not major problems. It was only in the 1990s, in fact, that many World War II veterans, then nearing the end of their lives, were inclined to talk about the war at all.[54] In the years immediately following the war, they were focused on getting an education, a job, a spouse, a car, and a house.[55] Their recollections a half-century later rarely focused on their own heroism, broad strategies, or tactical goals. Most were personal anecdotes about survival and about other soldiers—a type of narration usually too individualized to be heavily influenced by outside media. I concluded that anecdotal oral history accounts obtained more than fifty years after the event are useful in providing a greater understanding of the battle of Okinawa, and also that, with proper comment on context, oral interviews of low- and mid-level combat personnel helps recapture the experience of that battle.[56]

The Narrators

Of the more than twenty Okinawa veterans I interviewed, primarily in the 1990s, I selected six as primary narrators for this account. Five were artillerymen and one was a primary consumer of artillery support, an infantry company commander. Five of the six were junior officers. All six were motivated and intelligent. Most had at least some college education at the time of the war and acquired more following its conclusion. Most held white-collar jobs after the war. Although oral history is often considered "history from below" (i.e., from the lower classes, who rarely write their own history), this account is predominantly "history from the middle." One of the primary accounts, however, is that of a private, first class, and additional views of the enlisted men are contained in some correspondence. The oral accounts are supplemented with excerpts from the official records and the official U.S. Army account of the campaign in order to provide command-level insights and precise chronology. I also quote from the private letters of General Buckner to his wife.

Capt. Charles Sheahan

The highest-ranking artillery narrator is Charles Sheahan, who was promoted to captain (from first lieutenant) midway through the Okinawa campaign. During the course of the campaign he acted as an artillery liaison officer to battalion-sized infantry units, although he also operated in support of the 2nd Battalion. He was assigned to the Headquarters and Headquarters Battery of the 361st Field Artillery Battalion of the 96th Infantry Division.

Sheahan was born in 1920 and grew up in Brooklyn. He belonged to the Army Reserve before Pearl Harbor. He served with the 96th Division during the operations against the Japanese on the island of Leyte in the Philippines that began in late 1944 prior to Okinawa. After the war he went back to the New York City area and used the GI Bill to finance his college education. He became a junior high school teacher of physical education, science, and math. He performed another stint of active combat service during the Korean War and then continued teaching in the New York City region until retirement. He and his wife then moved to Florida, where he was still serving as a substitute teacher when I interviewed him in 1999. He died in 2009.

Lt. Ray D. Walton Jr.

Ray Walton Jr. was a second lieutenant of artillery during the Okinawa campaign. He was assigned to B Battery of the 361st Field Artillery Battalion, where he served as a forward artillery observer. Walton worked almost exclusively in support of smaller company-level units of the 2nd Battalion, specifically F Company and G Company.

Born in 1921, Walton graduated from Oregon State University in 1943 with a degree in chemical engineering. He was inducted in March 1943 while he was still a student. Although Walton had been enrolled in the Reserve Officer Training Corps (ROTC) for four years, the Army required him to attend Officer Candidate School at Fort Sill. Following his training there, he married, served briefly in Alabama, and then sailed from San Francisco in December 1944. Lieutenant Walton joined the 96th Division at Leyte, when his unit was in its final gun position of that campaign. After the war, Walton used the GI Bill to finance a graduate degree in chemical engineering. His civilian career involved him in the production process for nuclear weapons and the management of waste products from that process, first with General Electric Company and then with the U.S. government. Following retirement, he lived in the Maryland suburbs outside Washington, D.C.

Pfc. Charles P. Moynihan

Charles Moynihan was with the 361st Field Artillery Battalion, 96th Infantry Division during the Okinawa campaign. He served primarily as the artillery liaison team radio operator for the 2nd Battalion of the 381st Infantry Regiment.[57] He worked under the direction of the battalion liaison officer, usually a captain (among them Captain Sheahan). During the Okinawa campaign, Moynihan typically remained in close proximity to the infantry battalion leader, often not far behind the front line.

Moynihan was nineteen years old when he joined the Army Reserve in 1942. At that time he was attending junior college in Modesto, California. The government activated his unit in June 1943, and he received artillery basic training. He was selected for the ASTP (Army Specialized Training Program), which was intended to prepare technical personnel for service in Europe. While still in the Army he attended the University of Illinois and later the University of Wisconsin. When the Army closed the ASTP, Moynihan was assigned to the 96th Division.[58] He served on both Leyte and Okinawa. After the war, he graduated from the University of California at Berkeley and worked for Sears Roebuck in retailing. When I interviewed him in the late 1990s, he lived in California. He is now deceased.

Lt. Alfonse DeCrans

Alfonse DeCrans served as a forward observer during the Okinawa campaign. Like Walton, he typically provided fire support for infantry companies belonging to the 2nd Battalion of the 381st Infantry Regiment. DeCrans grew up on a farm, and that background came in handy on Okinawa when some American soldiers captured a five-hundred-pound steer and called on DeCrans' butchering skills.[59]

DeCrans started college at North Dakota State College in September 1939 and joined the Army National Guard the next month. He took ROTC courses for one full academic year plus two additional quarters. In April 1941 his National Guard unit was federalized and called up for active duty. By that time DeCrans was a sergeant in his artillery outfit.[60] DeCrans wanted to become a fighter pilot but was rejected because he was color blind. He served as a drill instructor in the 188th Field Artillery Regiment at Fort Lewis, Washington. Although he had reached the rank of staff sergeant, he decided to attend Artillery Officer Candidate School at Fort Sill, Oklahoma. There he came across an earlier acquaintance—Donald "Buck" Burrill.[61] The Army careers of DeCrans and Burrill—both narrators in this account—frequently overlapped. Both men served with the 27th Division in Hawaii before volunteering for reassignment. DeCrans arrived in the Philippines in late October 1944, a few days after the 96th Division had landed on Leyte. After interviews with Brig. Gen. Robert G. Gard (the division artillery commander) and Lt. Col. Avery Masters, he was assigned to the 361st Field Artillery Battalion, B Battery, where he served as the executive officer. After the war, DeCrans owned and operated a family farm in Minnesota. In 2009 he was living in Fargo, North Dakota.

Lt. Donald "Buck" Burrill

Donald Burrill served as a forward observer during the Okinawa campaign. He typically supported B Company of the 1st Battalion, 381st Infantry Regiment, 96th Division, although on occasion he worked in support of other companies belonging to the 1st Battalion. Despite their relatively junior rank, Burrill and DeCrans had been in the U.S. Army and in the Pacific for some time before the battle of Okinawa.

Burrill (1920–2005) hailed from Faulkton, South Dakota. Like DeCrans, he served in the 188th Field Artillery Regiment at Fort Lewis, Washington, where they met. DeCrans had served as a drill instructor; Burrill stood at the bottom of the U.S. Army hierarchy as a trainee. They went to Artillery Officer Candidate School at Fort Sill around the same time, although Burrill actually started the course before DeCrans.[62] Like DeCrans, Burrill served with the 27th Division in Hawaii earlier in the war, before the Saipan (mid-1944) and Okinawa campaigns (1945). The 27th Division was primarily made up of people from the New York City area. Burrill and DeCrans, concerned about the proficiency of that unit, volunteered for assignment elsewhere.[63]

Like DeCrans, Burrill was eventually sent to the 96th Division, which was composed largely of draftee farm boys from the Midwest. They could drive, shoot, and fix things. He liked the 96th Division much better. They were "darn good people."[64] Burrill arrived in the Philippines in late October 1944, a few days after the 96th Division had landed on Leyte. He underwent the same interview process in the 96th Division as DeCrans did and was assigned to A Battery of the 361st Field Artillery Battalion.

Burrill stayed in the Army Reserve after World War II and was promoted to major. Had he been willing to spend six months attending the Command and Staff School at Fort Leavenworth, Kansas, he could have been a reserve lieutenant colonel, but time constraints forced him to decline.[65] In the late 1990s Burrill lived in Casper, Wyoming. He is now deceased.

Capt. Willard G. Bollinger

Willard Bollinger provided insight into artillery's successes and failures from an infantryman's point of view. He commanded Company F during the Okinawa campaign and was among the 2nd Battalion company commanders who worked with Walton.[66] Nicknamed "Wild Bill" Bollinger by some of his men, he had a walrus-style mustache and wore his fatigue cap cocked to the side.[67] He was known as an aggressive frontline leader. As the Japanese would find out on Hacksaw Ridge, Bollinger sometimes did the unexpected.[68]

DeCrans, who dealt with the company commanders rather than the battalion commanders, considered Bollinger the "tops of any soldier" he ever saw and remembered him as an athletic, "well-built guy"—not someone that you wanted to get into an argument with. DeCrans also considered Bollinger a gentleman.[69] Bollinger was a good friend of Capt. Louis Reuter Jr., who commanded G Company, also in the 2nd Battalion of the 381st.[70] Bollinger and Reuter had gone to Officer Candidate School together and had both arrived as brand-new second lieutenants at Fort Adair, Oregon, when the 96th Infantry Division was being formed in 1942. Both were eventually promoted to company commanders. After the war Bollinger worked as a pension benefits specialist in the insurance industry for forty-five years.[71] In the late 1990s Bollinger lived in St. Louis, Missouri. He is now deceased.

Other Sources

SSgt. Karel Knutson

Karel Knutson served only one three-day stint on a forward observation team on Okinawa and is not one of the six primary narrators. His recollections are nonetheless occasionally helpful in providing the perspective of the men who actually fired the artillery pieces on the targets selected by the forward observers. An artillery sergeant, Knutson (born ca. 1922) spent most of his time operating the howitzers themselves. Prior to his Army service he had been a laborer, worked in a Civilian Conservation Corps camp, and operated heavy equipment. Knutson spent all four of his years in the Army (1942–46) with B Battery of the 361st Field Artillery Battalion, beginning as a cannoneer and working his way up to staff sergeant and chief of section. He was in charge of one of the battery's four 105-mm howitzers.[72] In 1999 Knutson lived in Minnesota.

Pfc. Don Dencker

Don Dencker was not a forward observer on Okinawa and does not serve as a narrator for this account. He was not interviewed for this book. Nonetheless, he indirectly influenced this account in several ways.

Dencker wrote the only published 96th Division infantryman's memoir of both the Leyte and Okinawa campaigns. A mortar man in an infantry company of the 382nd Regiment, Dencker fought through the entire Leyte and Okinawa campaigns. Like Bollinger's recollections, Dencker's book occasionally provides an infantryman's perspective on field artillery support. Dencker generously donated his time for a critical review of an earlier draft of this account. Dencker served several terms as the historian of the 96th Division Association and acted as my tour guide during a 1995 trip to Okinawa.

CHAPTER 1

INVENTING THE AMERICAN MOBILE ARTILLERY OBSERVER

For centuries artillery depended primarily on direct (line of sight) fire. During World War II, however, small teams of forward observers (FOs) positioned themselves with the infantry or armor along the line of battle. Typically led by an artillery lieutenant, they relied on binoculars, maps, radios, and field telephones to report enemy targets back to artillery batteries located some distance to the rear. Described as "probably the single most effective killer in ground combat," they were central to the success of the U.S. Army's efforts to dislodge and destroy the enemy once the battle on the ground was joined.[1] In spite of observers' prominent role, however, no history exists of the American forward observer. Indeed, at least one commentator claimed that military historians have largely ignored field artillery tactics on the grounds that they are "entirely technical and unsuitable for the general reader."[2]

"Prior to World War II," one American military authority noted, "the field artillery had no forward observers working with maneuver elements."[3] The observers who fought with Allied forces in World War II can trace their existence to innovations developed barely a decade earlier (1929–41). The prime innovators for this development were a handful of resourceful American field artillery officers largely based at Fort Sill, Oklahoma.[4]

During the American Civil War (1861–65), the opposing armies positioned their artillery in close proximity to the infantry. The gunners could usually see their targets. Under this system, each individual gunner was responsible for firing his own gun. If forward observers were used at all, it was generally only to spot the enemy's movements.[5] If the gunner could not see his target, someone had to go forward and signal the location back to him, usually with signal flags or arms.[6]

At the end of the nineteenth century the advent of indirect fire—aiming at targets not visible from the gun position—greatly increased the need for forward observers.[7] The increased range of artillery during the Franco-Prussian War (1870–71) began to demonstrate both the desirability and the capability of hitting targets outside the sight of the gunner.[8] The guns increasingly needed to be hidden from enemy small arms fire and field artillery fire, further limiting the gunner's view.

In 1882 Russian officer Karl G. Guk studied the ramifications of indirect fire. Guk, who had been a siege artillery commander during the Russo-Turkish War (1877–78), was an experienced artilleryman who recognized the potential role of a forward observer in the indirect fire scheme. He was perhaps the first person to articulate this concept and its corollary—that forward observers themselves would become the enemy's prime target once the firing batteries retreated out of sight.[9]

Experimentation with indirect fire increased as the end of the century approached. During the 1890s the British occasionally concealed artillery in positions behind hillcrests. In 1897 German artillery theorist Major General Moritz Elder von Reichold went so far as to assert that to survive on the battlefield, field artillery had to limit itself to indirect fire. Japan's success with indirect fire during the Russo-Japanese War (1904–5) tended to support von Reichold's argument, even though the technique was not immediately widely adopted. American experimentation began about this time as well. By 1896 U.S. War Department doctrine had acknowledged that it was helpful to place an officer well in advance of the guns to note the impact of the artillery and its coordination with friendly infantry. Innovators thus sought to improve on the signal flag method of communication. Tactical experimenters attempted to link forward observers to the artillery batteries by means of field telephones. By 1912 each American artillery battery had three telephones.[10] In the early years of the twentieth century, the forward observer assumed a more active role in the direction of field artillery fire.

During World War I (1914–18), artillery was the dominant combat arm. Although World War I is more often associated with machine guns and gas warfare, artillery shell fragments inflicted an estimated 70 percent of the wounds suffered during that conflict. Nonetheless, artillery had been unable to live up to the claims of its advocates that "artillery conquers and infantry occupies."[11] Noted military historian John Keegan summed up the problem as follows: "What had not been perceived is that firepower takes effect only if it can be directed in timely and accurate fashion. That requires communication. Undirected fire is wasted effort, unless observers can correct its fall, order shifts of target, signal success, terminate failure, co-ordinate the action of infantry with its artillery support. . . . Nothing in the elaborate equipment of the European armies of the early twentieth century provided such facility."[12]

The unexpected dominance of indirect fire in World War I sparked a military revolution.[13] Even the quite advanced British artillery had begun the war in 1914 with a preference for firing over "open sights." If British artillery officers wanted binoculars to observe fire in 1914, they were expected to purchase them at their own expense.[14] The British suffered as a result of their retention of traditional tactics early in the war. At the battle of Le Cateau in August 1914, the British artillery commander deliberately elected to deploy his artillery for direct fire use rather than indirect fire operations, and it cost them dearly.[15] The British remained dependent on field pieces such as 18-pounders, which tended to fire inaccurately when used in an elevated (indirect fire) mode. Conversely, the Germans planned (and equipped themselves) for indirect fire operations. The German artillery had a much higher percentage of howitzers than the British—howitzers, with their high-angle trajectory, are designed for use in an indirect fire mode—and German artillery triumphed in the struggle at Le Cateau. Shelford Bidwell noted that "7 of the 12 [British] field batteries were beaten . . . into silence, and 26 of the 42 exposed guns were left in the hands of the enemy."[16]

The devastating effect of German counter-battery barrages soon caused the British army to reconsider indirect fire.[17] Indirect fire, however, gave rise to another problem. The gunners could not hit what they could not see.[18] As a response, the British developed forward observers and began to use observation posts (OPs) to adjust artillery fire. Aircraft continued to carry out artillery spotting missions, but more commonly the direction and adjustment of fire took place in hillside dugouts and other OPs.[19] Use of telephones combined with the much longer range of modern artillery allowed artillery officers in OPs to call in fire from guns several miles away. Unfortunately, Russell Gugeler noted, the light telephone wire connecting the liaison officer to the distant firing batteries "failed more often than it functioned."[20] Telegraph lines suffered from the same liability, and relay runners could be cut down by enemy fire. The British gunners were thus often forced to rely on "unobserved indirect fire."[21]

French and British observers were horrified by American doughboys' lack of technical artillery skills. The Americans, after all, had enjoyed the luxury of three years of peace in which to learn from the errors of the belligerents. American deficiencies were particularly striking in the areas of artillery method and cooperation between different combat arms.[22] High-ranking American officers, including Maj. Gen. W. J. Snow, the chief of artillery, agreed: "The condition of the field artillery as regards its organization, its equipment, its training . . . was nothing short of deplorable and chaotic."[23]

World War I methods simply could not supply flexible and accurate support for the infantry. In particular, the fire could not be rapidly shifted around the battlefield. Even with observers and indirect fire, field artillery failed to provide adequate direct support to the infantry.[24] At the battle of Soissons (the Second Battle of the Marne, July 1918), for example, American colonel L. Upton reported that "there was absolutely no control of artillery fire by the troops in line or by the commanders" in the 9th Infantry Regiment.[25] At Saint-Mihiel in September 1918, American gunners ignored an order from the American commander, Gen. John J. "Black Jack" Pershing, that put field artillery under the control of infantry brigades.[26]

Efforts to coordinate infantry actions with artillery fire all too often resulted in friendly fire casualties. A high-ranking French officer asserted that errant French artillery had killed 75,000 of his nation's infantrymen. Historian David T. Zabecki reported that "some units in the French army even resorted to sewing large white patches on the backs of their uniforms in the hope it would prevent their own guns from firing on them."[27]

The static World War I forward observer was very much part of the problem. James Russell Major, an American World War II forward observer (and later a writer of European history) criticized his Great War predecessors for having "no direct contact with the attacking infantry." Gugeler noted that "usually artillery liaison officers worked with infantry commanders to plot the locations of desired barrage fire, but there were neither artillery observers who could see the fire nor a reliable means of communicating with the firing battery except by messengers either on foot or horse, or less often the use of semaphore."[28] The lack of communication presented a major problem for military tacticians. Even staunch advocates of artillery admitted the limitations of the big guns. A 1921 British training manual bluntly conceded that "artillery cannot ensure decisive success in battle by its own destructive action. It is the advance of the infantry that alone is capable of producing this result."[29] But the infantry needed close artillery support to achieve that success, and such support was lacking.[30] One critic compared "artillery without proper observation [to] blind Polyphemus . . . throwing rocks in the hope of hitting one of his tormentors."[31]

The experience of the Great War brought American gunnery instructors to the realization that artillery had to find a better way to assist the infantry.[32] Although European powers had just endured four grueling years of trench warfare on the Western Front, the post–World War I U.S. Army elected to focus on open warfare rather than trench warfare. A few farsighted American officers envisioned a more mobile battlefield made possible by continual improvements in motor vehicles. Thus the Americans sought to make rapid infantry advances supported by artillery fire routine. This concept required both that field artillery be capable of rapidly shifting fire around the battlefield and that the gunners maintain constant communication with both infantrymen and the recently developed armored vehicles.

The initial American experimentation with forward observers began seriously in the 1920s at Fort Sill's Field Artillery School, where the U.S. Army had established a base for the intellectual examination of the art of gunnery.[33] Theorists at the school looked for ways to improve on World War I artillery tactics. Maj. Jacob Devers, head of the Gunnery Department, "trained forward observers and experimented with letting them control the battery fire from positions near the combat zone."[34] At the time, artillery organizational specifications still did not include an observer to accompany the infantry.[35]

Maj. Carlos Brewer, who replaced Devers in 1929, continued the experimentation. Rather than each field artillery battery firing independently, Brewer tried to centralize the targeting information and firing calculations at the level of the artillery battalion. This was the origin of what eventually became known as the "fire direction center," or FDC. On being notified of an enemy target, the men in the FDC would calculate the "technical gunnery solution"—that is, instruct three separately located batteries how to direct their guns against a specific target. Precise targeting could be done in one of two ways. Either a target survey could be conducted, which would take two to six hours—too long for Brewer—or forward observers could communicate directly with the field artillery units. Only with such observers could the artillery attack targets of opportunity quickly and accurately.[36]

The innovators at Fort Sill changed the way observers communicated with the gunners. The state-of-the-art method for calculating target information, borrowed from the French during the Great War, was both rigid and mathematically complex. American field artillery theorists of the 1930s sought to simplify the system. In order to avoid lengthy discussions about the target location, both the forward observer and the field artillery battalion used maps with pre-designated base point locations. When calling in a "fire mission," the observer could tell the gunnery unit where he wanted fire in relation to a prearranged location. The observer's report of "one hundred right, two hundred under," for example, would tell the gunners to fire one hundred yards to the left of the location and to increase the range by two hundred yards.[37] The simplified system was easy to learn, and that would be important in places such as Okinawa, where the high casualties suffered by forward observation teams resulted in the rapid turnover of personnel.

As the U.S. Army reconsidered the role of artillery, the innovators found inspiration from *Field Guns in France*, written by British field artillery officer Lieutenant Colonel Neil Fraser-Tytler in 1929. Fraser-Tytler insisted that an artillery observer should accompany the infantry even during attacks. He described experiments during World War I in which he had been able to shift artillery fire around the battlefield and thus reduce enemy strong points that were holding up the infantry. He accomplished the feat by using a telephone line laid out as he advanced. Fraser-Tytler's work gained currency at Fort Sill. Maj. Orlando Ward, who succeeded Brewer as head of the Fort Sill Gunnery School in 1932, admired it. Between 1932 and 1934 officers from the school's Gunnery

Department developed and tested the FDC concept. The Army did not adopt this innovation officially until 1941, and the FDC would prove central to artillery operations during World War II.[38] Some scholars even maintain that the FDC, rather than forward observers, was the key to solving the indirect fire problem that had risen during World War I.

Major Ward further proposed that the FDC arrange the targets in order of importance. Thus the priorities of the overall commander, not those of the artillery unit or the forward observer, determined the order in which missions would be fired. Moreover, the FDC had to consider the safety of nearby friendly units. These decisions, taken in the 1930s, would have important consequences during World War II. Forward observers (traveling with the infantry and armor units) might not always receive the fire support they had requested. The experience of 1st Lt. Oliver J. Thompson, a field artillery liaison officer supporting the 96th Division's 382nd Infantry Regiment, illustrates the point. In late October 1944 he was on Leyte supporting an infantry battalion advancing on a Japanese stronghold in the village of Tabontabon. An adjacent infantry battalion had already engaged in a firefight there. Thompson's battalion, forming the left wing of the American advance, came to a halt. He believed that he had seen two Japanese soldiers in a ditch ahead of him and radioed in a request for artillery fire. "I was refused," he later recalled, "and at the time I didn't know why." Soon afterward his infantry battalion suffered casualties, including Thompson himself. The next day, while sitting "on a jeep pulling a trailer with dead bodies," the wounded and disgruntled lieutenant ran into the commander of his artillery battalion. When Thompson asked him about the refused fire mission, the colonel "explained that the 2nd Battalion was approaching *toward us* on the same road."[39] Observers thus occupied a delicate position because they often had the job of telling the maneuver units that no artillery support was forthcoming.[40] Although they were simply the messengers and not the decision makers, the observers undoubtedly tasted the wrath of the ground forces whose pleas for assistance went unheeded.

The new artillery tactics developed at Fort Sill during the 1930s would pay enormous dividends in the next war. This work took place during the crisis of the Great Depression (1929–39). The economic crash affected the Army in various ways, but gunners limped along with a bare bones budget. Historian Larry Roberts noted that the Field Artillery School had "to borrow an automobile from a local car dealer in order to conduct motor maintenance classes."[41] President Franklin Roosevelt's administration initially focused the Army's attention on national economic recovery rather than national defense.

Soon after Roosevelt took office in 1933, the Army was tasked with administering the initial aspects of the Civilian Conservation Corps (CCC), a New Deal program to put unemployed young people to work in national parks, forests, and public lands. The Army had to cut back on its schools in order to process the massive numbers of youths who leaped at the chance to earn some money.[42] The Field Artillery School at Fort Sill, which normally closed in June for a summer break, closed a month early in 1933 and sent some 60 percent of its faculty to help in the CCC camps. Fort Sill was, however, able to reopen on schedule in September 1933.[43]

Despite these distractions, artillery intellectuals continued their work. Technological improvements in the radio proved decisive. In 1939, for the first time, American artillerymen accompanying ground troops had man-portable radios available to them, freeing them from telephone lines and giving them the mobility they needed to follow the "ebb and flow" of battle.[44] The U.S. Army and artillery nevertheless remained well behind the developments taking place in the German army as war broke out in Europe. In 1939 the German army introduced the mobile forward observer concept into actual combat operations. In Poland, the German Wehrmacht attached radio-equipped field artillery observers to the armor and infantry spearhead of their blitzkrieg. At the same time, the pre–World War II organization of American artillery units remained wedded to static observers. Maj. Gen. David Ott, an important late-twentieth-century artillery theorist, noted that "each battery of field artillery had a reconnaissance officer whose duties included the establishment of a battery observation post. This post was then manned by either the reconnaissance officer or the battery commander and all observed fires were conducted from the battery observation post by one or the other of these two officers. Traditionally observed fires were fired by only one battery."[45]

While the United States remained a nonbelligerent, its Army artillery leaders quickly grasped the importance of the German combined-arms approach, which they reproduced during peacetime maneuvers conducted in the southern United States in 1940 and 1941. The exercises demonstrated the ability of ground observers to direct large amounts of artillery fire onto important targets—a significant advance. In October 1941 the chief of field artillery gave final approval to the FDC concept and the new artillery techniques it encompassed.[46] The Americans were quick to claim that the "technique of massing artillery fires quickly and accurately is a genuine American first."[47]

To accompany the new technical developments, the U.S. Army created a new organizational scheme to match artillery with infantry (or armor). Ott described the strengths and limitations of the scheme:

> During World War II the concept of an observer with a maneuver company was perfected. However, our table of organization and equipment was extremely austere. Each firing battery had only one forward observer and, of course, had to support three maneuver companies as a minimum. To do this, the battery reconnaissance officer was traditionally used as an additional forward observer. In many cases observers moved from one company to another as they passed in reserve or were committed so that the direct relationship of observer to company was not possible. We also used assistant executive officers from the firing battery, riding in a vehicle usually scrounged from the wire section to provide for an additional FO. It was difficult to work but, one way or another, we generally had a forward observer with every company in contact.[48]

With the new organizational structure, new communication capabilities, and new technical advancements, field artillery observers came into their own during World War II. Radio-equipped American forward observers operated with infantry units on Guadalcanal and in North Africa in 1942 and 1943.[49] By at least early 1943, the new methods of adjusting fire were filtering down to American college students enrolled in ROTC. During 1944–45, American forward observers achieved an enviable record in Western Europe. Some historians even maintain that the Germans had little respect for American infantry but much admiration for American artillery.[50]

The interwar artillery developments clearly constituted a grand success for the U.S. Army. By 1944 one American writer asserted that Fort Sill had become the "greatest field artillery school in the world."[51] A distinguished artillery historian called the FDC and man-portable radios "the most striking difference between World War II and World War I field artillery capabilities."[52]American prowess in gunnery was an important development because, in the eyes of some observers, at least, artillery dominated all other arms on World War II battlefields. In his memoir, James Russell Major noted that "half the battle casualties were caused by artillery fire, in spite of the commonly held view that tanks and air power were the dominant weapons."[53]

The World War II forward observer stood out as the critical factor in the improvement of American artillery. In a defensive role, a single forward observer from the 30th Infantry Division was instrumental in halting the German counterattack on Mortain in August 1944. Although the enemy assault cut off his battalion from other American units, Lt. Robert Weiss managed to radio in 193 fire missions in 6 days—an average of one every 45 minutes.[54] Later in 1944, forward observer Lt. Howard Kettlehut adjusted the fire from 18 battalions of artillery in Germany's Hürtgen Forest in an effort to save a U.S. Army Ranger unit from destruction. World War II historian Stephen Ambrose concluded that "American artillery saved the men."[55] In an offensive role, fast-moving American armor forces attempting to relieve Bastogne during the battle of the Bulge could rely on their artillery support to keep up with them. In order to blast through Assenois, the last significant German stronghold blocking his path to Bastogne, Lt. Col. Creighton Abrams famously radioed back for decisive artillery fire with the words "Concentration Number Nine; play it soft and sweet."[56] By 1945, American artillery had finally surpassed European artillery in terms of sophistication.[57]

CHAPTER 2

PRELUDE TO OKINAWA

America's entry into the Pacific war was sudden but not entirely unexpected. U.S. Navy planners had been considering the possibility of war with Japan since the Russo-Japanese War of 1904–5. American planners continued to discuss and revise their plans right up until the outbreak of World War II in Europe. In 1940–41 the first war plan, code-named Orange, was officially replaced by Rainbow 5, a worldwide plan of U.S. military operations. The concepts of Orange nonetheless remained highly influential in the actual Central Pacific campaign. The contingency plan correctly assumed that the Japanese would attack American forces in the Philippines early in the war. In response, the first phase of the plan called for the U.S. fleet to assemble at Pearl Harbor. The middle phase called for the American armada to fight its way to the Philippines. The final phase of the plan proposed the seizure of one or more islands in the Ryukyu chain just south of Japan for use as a blockading base against Japan. The largest and most suitable island in the Ryukyus was Okinawa. Planners assumed that Japan could not yield the Ryukyus and survive. They anticipated ferocious Japanese resistance.[1]

American planners also assumed that the Japanese navy would eventually be forced to fight a Trafalgar-like decisive battle against the American fleet. Without Western Pacific repair facilities, the United States could defeat Japan in every battle but still lose the war. The Japanese could repair their ships at home, whereas the Americans would lose ships permanently by attrition. Thus the Orange planners deemed seizure of Western Pacific dockyards—particularly in the Philippines—critical. Only the middle phase of the American plan—moving the U.S. fleet from Pearl Harbor to the Philippines—raised any significant dispute among prewar planners. Aggressive planners advocated risking a nonstop "through ticket" drive to the Philippines by the American fleet. The more cautious called for establishing intermediate bases in the mid-Pacific, correctly believing that forward support bases would allow the U.S. Navy to operate at a tempo the Japanese could not match.[2]

The campaign against the Japanese after Pearl Harbor followed War Plan Orange but with significant modifications. First, the United States made a two-pronged attack across the Pacific rather than a single thrust. Adm. Chester Nimitz led the Central Pacific prong. Nimitz's planners had been trained on Plan Orange.[3] Nimitz himself believed in a step-by-step advance rather than the "through ticket" drive to the Philippines.[4] Thus his Central Pacific forces would seize Tarawa, Kwajalein, Eniwetok, and Saipan one by one. That advance across the Central Pacific consumed two and a half years (from mid-1942 through the end of 1944).[5]

The southern prong of the attack advanced across the Southwest Pacific under the direction of Gen. Douglas MacArthur. The Southwest Pacific campaign, unlike the Central Pacific campaign, had not been the subject of prewar planning and war games.[6] American planners had long assumed that the United States would have to fight Japan without any allies. As World War II proceeded, however, Australia added both positive (an unexpected ally with capable ground forces) and negative (an unexpected defense obligation) features to the U.S. plans. Both prongs of the Pacific attack contributed to the eventual American victory, but most historians count the Central Pacific campaign as the primary and decisive thrust.[7]

The two prongs assisted each other in the Philippine Islands. General MacArthur undertook the primary task of seizing the archipelago, but forces from Nimitz's command helped in the operation. Prior to the outbreak of the Japanese-American conflict, the Philippines had been an American possession (scheduled for independence in the 1940s). The Japanese seized control of the islands in 1942. As he was forced out of the Philippines, MacArthur gave his famous pledge: "I shall return."[8] His return, however, did not occur for more than two years.

The recapture of a western naval base somewhere in the Philippines had been a critical element in the various permutations of Plan Orange, although seizure of the Philippines' major harbor—Manila Bay on Luzon—was not required for military purposes. Other islands, such as

Leyte, could also provide the necessary platform for the final phase of Plan Orange. In particular, MacArthur sought to exploit Leyte as a logistics depot and air base. Dumanquilas Bay on the island of Mindanao was for a time a candidate for the key American naval base in the Philippines, but on September 13, 1944, the Joint Chiefs of Staff ordered MacArthur to bypass Dumanquilas Bay and seize the more northerly island of Leyte instead.[9] His forces struck there in October 1944.

The 96th Division saw its first combat as one of MacArthur's assault landing divisions on Leyte, and the division's forward artillery observation units had their first experience directing fire on an enemy there. Charles Sheahan saw his first combat at Leyte. Sheahan had been assigned to the 96th Division as a lieutenant when it was first activated at Fort Adair, Oregon, in August 1942. The Deadeyes—nicknamed for the division's emphasis on marksmanship—received further instruction after sailing to Hawaii in July 1944. In Hawaii they reembarked for the voyage that concluded in the Philippine Islands.[10]

On October 20, 1944, the 96th Division made an amphibious landing on the Leyte beaches as Japanese and American aircraft dueled in the skies overhead.[11] A bullet that seemed to have originated from the dogfight fatally struck a B Battery soldier as he took a break to eat. SSgt. Karel Knutson commanded one of the howitzers of B Battery in the 361st Field Artillery Battalion. He remembered the tough fight for Catmon Hill early in the Leyte campaign. During the first nine days on Leyte his battery operated twenty-four hours a day and no one in his section slept. After that, his men slept in two shifts. They fought for 112 days. Not even a successful Japanese air strike on an American ammunition dump managed to slow the American artillery effort in any meaningful way.[12]

Following the successful landing, the Deadeyes sought to secure and expand the American beachhead. Since so much of the inland terrain was mountainous jungle, the seizure of the primitive road network became a critical objective. The task of opening and securing a road was typical of those assigned to 96th Division infantry battalions. It was in this environment that Lt. Oliver Thompson gained his first combat experience. Neither his 362nd Field Artillery Battalion nor the 382nd Infantry Regiment units that he supported had the maps they needed. Circumstances thus dictated unusual procedures. "In the jungle," he recalled, "the forward observer would call for smoke to locate the road. Then he would adjust the high-explosive rounds by sound." Even sixty-five years later, the memories were fresh in Thompson's mind. Above all else he remembered the stress forward observers felt in the jungle. He recalled that at one point the men "were in some kunai grass on our stomachs because the Japs pinned us down with machine-gun fire. To stand up would be fatal. I told my radio operator to raise the aerial a little to see if he could get battalion [the FDC]. Luckily he succeeded. Then I crawled on my hands and knees to a tree and stood up and shouted my adjustments to the . . . radio operator." The foot soldiers appreciated his efforts and rewarded him for his skill. "I was put in for a Bronze Star by the infantry," Thompson later wrote.[13]

As they would on Okinawa, the 361st supported the 381st Infantry Regiment. Lieutenant Sheahan served high in the Leyte hills, and he later compared warfare on Leyte with the guerrilla fighting in Vietnam. His equipment included a cumbersome fifty-pound Model 110 radio that was intended to be moved in a jeep. "[On Leyte] there wasn't a front. . . . [O]n Okinawa there was a front, which makes it a lot easier. . . . On the Philippines it was more like a Vietnam guerrilla type of thing. A group over here, a group over there. There was no definite front. You go up into the mountains and maybe you run into somebody and maybe you don't."[14]

One day up in the mountains, Lieutenant Sheahan directed fire against a Japanese knee-mortar position. A knee mortar is a very small mortar easily carried by infantrymen. He could not see the mortar position. Just as Thompson had done, he directed artillery fire against the mortar position by using sound alone.

> You talk about accuracy? This is interesting. I never forgot it. In the Philippines we were up in the mountains up there. . . . So a fellow with a knee mortar—a Jap. In the woods and everything but he has a knee mortar. So I asked for one gun. One gun. They would blow it up. "Move it a little to the left. Move it a little to the left." I hear that thing still making a little click. . . . We fired. Later on we moved through the woods. I came upon him. The shell—this was the best shot of World War II. The shell . . . went

> behind him and blew him out of the hole and he sat back in the hole holding the knee mortar. . . . It was adjusted by sound. . . . It was definitely a Japanese mortar man.[15]

The Japanese forces punished elements of the Deadeye division in return. They ambushed G Company (381st Regiment, 96th Division) on Leyte and killed the company commander. When the despondent men returned to camp, Capt. Louis Reuter Jr.—newly returned to the unit—greeted each man by name. Reuter had been the G Company commander back in the States but had been reassigned to transportation duties within the 96th Division. After the shock of the ambush, Reuter's return inspired the men of the company.[16]

The hardest fighting centered on the northern third of Leyte—the portion closest to Luzon.[17] By December 1944, however, elements of the 96th Division were engaged in patrols described by the division history as consisting less of fighting and more "of hard marches up and down steep trails, of leeches by the thousands, of dysentery, dengue fever and jungle rot, of laboriously packing supplies on tired backs." The Deadeyes suffered 376 officers and men killed, 4 missing, and 1,289 wounded. Another 2,500 fell victim to injury or illness. Despite their casualties, the Deadeyes eliminated an estimated 7,700 Japanese soldiers.[18] Even so, infantry rifleman Pvt. William Filter considered Leyte a "minor skirmish" compared with the brutal Okinawa campaign. The Japanese on Leyte lacked strong artillery; that would not be the case on Okinawa.[19]

Max Hastings, a prolific twenty-first-century military historian and frequent critic of Douglas MacArthur, declared Leyte "a worse defeat than the Japanese need have suffered, a more substantial victory than MacArthur deserved." Indeed, MacArthur had not anticipated such ferocious resistance. Some commanders could hardly force their men to advance. Hastings also concluded that at Leyte the American "high command was without flair, and many infantry units were slow." From a strategic standpoint, Leyte proved to be a disappointment for the Americans. The island never became a significant Army Air Force base—a major goal in seizing it—because, as Hastings observed, the "waterlogged plains were wholly unsuitable for intensive aircraft usage, and even for stores depots."[20]

At the campaign's conclusion, the Deadeyes nonetheless took pride in their performance. Gen. John R. Hodge, the corps commander to whom the 96th reported on both Leyte and Okinawa, assured the Deadeyes that they had compiled a "brilliant and enviable" record in the Philippines.[21] With the American victory the troops relaxed, resting for more hard fighting to come. Lieutenant Sheahan recalled that there was "a lot of waiting around" after the battle. "On Leyte, you—once it was all over—you started cleaning up your guns, have a few ball games, have drills, a little training, you get new people in, . . . and wait for the next operation."[22]

As the Deadeyes recuperated in the Philippines during late February and March 1945, a hellish battle raged hundreds of miles to the northeast. The 96th Division did not fight on Iwo Jima. Nor did the overwhelming majority of the other ground troops that would be involved in the Okinawa campaign. The battle, however, had an enormous impact on the outlook of America's leaders and fighters. Neither the biggest nor the most important of the Pacific battles, Iwo Jima became the most famous battle of the war against Japan.[23] The battle did not officially end until March 26, 1945, nine weeks after it began, and mop-up operations continued for weeks.

The fighting on Iwo Jima was vicious—as it would soon be on Okinawa. Admiral Nimitz was correct when he told the world that among the Marines on Iwo Jima, "uncommon valor was a common virtue."[24] The Japanese suffered far more deaths (the entire garrison of 21,000 less a few hundred prisoners) on Iwo Jima than the Americans (around 7,000), but the Americans—right up to the commanders—suffered more total casualties (dead plus wounded). Of the twenty-four U.S. Marine infantry battalion commanders who began the battle, seventeen were killed or wounded badly enough to require evacuation.[25] Even the sketchy information about Iwo Jima available to the combat troops assembling for the Okinawa operation must have made them highly apprehensive. As the Americans approached the four Japanese home islands, the fighting would become increasingly brutal.

The seizure of Okinawa (or a suitable alternative in the Ryukyus) had always constituted a major goal of American planners, who believed the Ryukyu chain to be the "decisive position of the entire conflict."

Operation Iceberg, the plan to seize Okinawa, anticipated the invasion of Japan—first Kyushu, tentatively scheduled for November 1, 1945, and then Honshu, tentatively scheduled for March 1, 1946. Indeed, the Japanese considered Okinawa one of the Japanese home islands, even though it was more than 300 miles away from the four main islands. To make an analogy, it was as if Hawaii were only 330 miles off the coast of California and some Asian nation was planning to move its fleet across the Pacific, seize the islands, and kill virtually all of the U.S. troops there.[26]

For the Deadeyes, participation in the Okinawa campaign meant transfer to a different theater of war. When the 96th Division invaded Leyte in October 1944, it had temporarily fallen under the control of General MacArthur. Now the 96th was switched from MacArthur's Southwest Pacific Command to Admiral Nimitz's Central Pacific Command.[27] They would be part of the new Tenth Army being formed under the command of Lt. Gen. Simon Bolivar Buckner Jr., the son and namesake of the Confederate general best remembered because his unconditional surrender of Fort Donelson gave Union general Ulysses S. Grant his first major victory. In his previous assignment, the younger Buckner had performed the daunting task of defending the vast Territory of Alaska from Japanese attack. He had overseen the construction of the famed Alaska Highway—the first road into Alaska from the United States—to defend America's northern outpost. Between 1940 and 1944 Buckner had been promoted from colonel to lieutenant general as the men under his command completed "the biggest and hardest job since the Panama Canal."[28]

Capt. Willard Bollinger was surprised when he first learned about the Okinawa campaign at an infantry company commanders' meeting on Leyte. His infantry company had just finished the grueling Leyte campaign, and he was short of both men and equipment. He was able to get some replacement soldiers, but getting supplies was a different story. Most logistical resources on Leyte belonged to the Southwest Pacific Command, and MacArthur planned to use them against the Japanese in the battle for the heavily defended Philippine island of Luzon. He was reluctant to give those supplies to troops from another command. This posed a significant problem for Captain Bollinger and other commanders. His infantry company was so short of weapons that two of his men did not have rifles at all, and others had to use rifles that needed replacing.[29]

Captain Bollinger was not the only one with supply problems. The Navy faced an enormous task in supplying the Okinawa invasion force. The logistical support for the assault echelon alone involved loading 183,000 troops and 747,000 tons of cargo into 430 assault transports and landing ships at 11 ports stretched over 6,000 miles, from Leyte to Seattle.[30] Then the fleet had to sail through hostile waters to Okinawa. U.S. Navy lieutenant (junior grade) Harold Scott's vessel left from the mid-Pacific island of Eniwetok. Scott, a radar and radio repair officer, hailed from Wallowa, Oregon. His vessel, the attack transport USS *Hendry* (APA 118), could carry up to 1,700 amphibious assault troops.[31]

The troop convoys and supply ships dreaded Japanese submarine attacks. They followed a zigzag course day and night, even though it increased the risk of collision. During World War I and the early part of World War II, German U-boats had taught the British and Americans that vessels stopped in the water were easy prey for enemy submarines. Stopping for rescue operations jeopardized the lives of the entire crew. The decision that ships would not stop to conduct rescue operations in enemy waters was difficult but unavoidable. During one wartime amphibious operation, Scott had witnessed the human consequences of this decision when the *Hendry* lost one of its passengers. An assault trooper who had been sleeping on deck accidentally rolled off the *Hendry* and into the sea. A small ship in the convoy asked for permission to stop and rescue the man. Permission was refused. The man overboard thus had to watch the convoy steam over the horizon without him.[32]

Lt. Ray D. Walton Jr., a relatively recent arrival in the Philippines, supervised a small portion of the loading operations on Leyte. The equipment had to be stowed in a particular arrangement for a well-executed amphibious operation.

> Most of our Field Artillery Battalion, 105 mm howitzers, DUKWS [amphibious trucks], trucks, supplies were on . . . an LST [Landing Ship, Tank] in which we put down a layer of three feet of field artillery ammunition, all in heavy boxes, and they layered lumber on top of that. Then the trucks, and guns, the

> 105-mm howitzers, were put on top of the ammunition. That was the way to get the most into the ship. It was critical that the ship be loaded correctly so that the right pieces of equipment could come off first for the landing.[33]

Many elements of the 96th Division, including Walton's field artillery battalion, embarked and began their voyage to Okinawa on March 13, 1945. Walton did not, however, sail on the vessel he had helped load. As the battery's newest officer, he was not scheduled to land with that unit's first wave. Walton and a detachment of surplus men from the battery were assigned to an attack transport (APA) that carried some members of the 381st Regiment—the infantry unit that Walton's artillery battery supported. An APA could not sail as close to the landing beaches as an LST, but it traveled much faster on the open sea. The LSTs, which typically moved at ten knots, thus left the Philippines about a week earlier than the speedy APAs, which could make seventeen knots.

The plan was for the various ships of the invasion fleet to arrive off Okinawa simultaneously. At the same time, the U.S. Navy did not want to expose the speedy APAs, their decks crowded with soldiers, to Japanese submarines any longer than necessary. Somewhere around March 20, Walton's APA began its sprint to catch up with the slower-moving vessels in the fleet. Walton's voyage to Okinawa was not uneventful. The ship ran into "a typhoon with gale force winds and very high seas," he remembered.[34] To add to their problems, the transport ship had to take evasive maneuvers because a Japanese submarine had been reported in the area. The erratic zigzags had an impact on the ship's hull that evoked concern among the passengers. "We would travel a hundred yards, turn sixty [degrees], go another hundred yards, and then turn back. . . . Every time we turned, the whole ship would shake. We wondered if the ship was going to last, but it did," he recalled.[35]

Walton's voyage took about ten days; the LSTs carrying his artillery battery took about seventeen. Walton recollected that the ships carrying the 96th Division "all got to Okinawa during the night of March 31, 1945."[36] Operation Iceberg was under way.

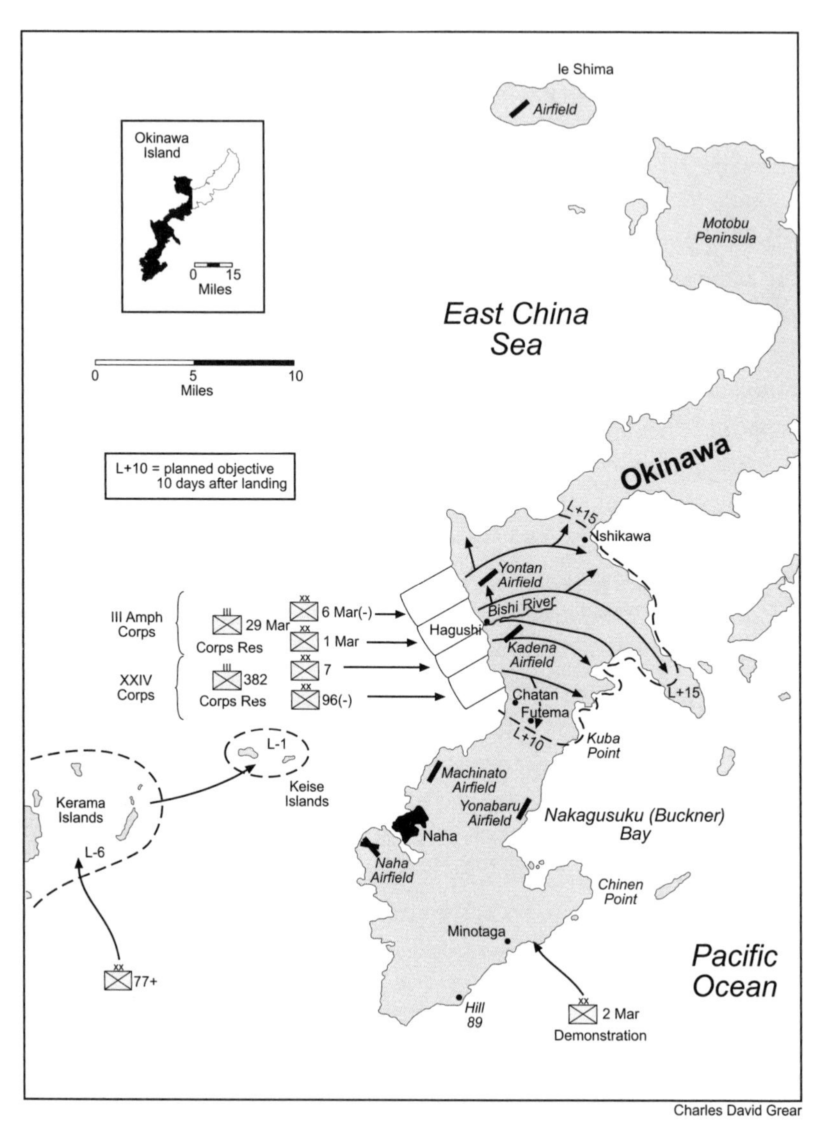

Map 4. The Invasion of Okinawa
Adaptation by Charles Grear based on map 3 in Roy E. Appleman, James M. Burns, Russell A. Gugeler, and John Stevens, *Okinawa: The Last Battle* (Washington, D.C.: Center of Military History, 1948, 1991).

CHAPTER 3

EASTER INVASION

The artillery duel between the Japanese and the Americans on Okinawa began before the first American soldier set foot on the island's beaches. Prior to the arrival of the main invasion fleet on the evening of March 31, 1945, elements of the U.S. Army's 77th Infantry Division had seized several small islands within sight of the main island of Okinawa and had established long-range heavy artillery on one of them—Keise Shima.[1] The Japanese were surprised, but they considered the long-range fire more of a nuisance than a threat and responded with counter-battery fire.[2]

The extent of the Japanese artillery response in turn surprised the Americans. Col. Bernard Waterman, the U.S. Army's corps artillery commander on Okinawa, later acknowledged that at Keise Shima the Americans had "received the first foretaste of a type of Jap artillery combat . . . unlike anything we had previously experienced—an appreciation of our own technique of massing fires."[3] During the two previous campaigns in which Colonel Waterman had participated, "Jap artillery had been a negligible factor."[4]

The invasion fleet that reached Okinawa consisted of 1,300 ships. The naval fire support alone consisted of 10 battleships, 9 cruisers, 23 destroyers, and 177 gunboats. The naval gunfire in support of the Okinawa landing was the heaviest concentration ever to support a landing of troops.[5] Lieutenant Walton watched as the "naval bombardment of Okinawa began, with 12-inch and 16-inch guns from battleships . . . plus aircraft carrier–based Navy fighter-bombers." Walton was impressed by the size of their shells when he later came across some unexploded duds on the Okinawa shoreline. "It was shocking to walk up on dud 16-inch projectiles on the beach area, see the size of the projectiles and know how much explosive they contained, and know that it was possible to explode at any time," Walton recalled. Fortunately, the U.S. field artillery on Okinawa had fewer problems with defective shells. "I never found a dud field artillery shell but found several 16-inch gun shells," Walton insisted.[6]

In order to tie down enemy reserves and keep the Japanese guessing about their intentions, the Americans staged a decoy landing in southern Okinawa many miles from the actual landing site. Maj. Roy Appleman, a historian accompanying the invasion forces, recalled that this "diversion simulated an actual assault in every respect."[7]

Lt. Harold Scott participated in the feint. The portion of the American fleet that included the *Hendry* arrived off Okinawa during the dark early-morning hours of April 1. Shortly after the vessel came to a halt, Scott noticed "a persistent faint signal on the surface radar screen" and "thought it might be a periscope from a submarine." The *Hendry* tried to inform the rest of the fleet, but the radio was not working. Checking the problem proved a risky undertaking because the voice radio antenna was at the tip of the yardarm, fifty to sixty feet above the rolling dark ocean. Scott had no choice but to climb the rear mast and "shinny out on the yardarm to disconnect the cable."[8] Straddling the yardarm, he inched himself out about fifteen feet, replaced a section of the cable that looked distorted, and was hugely relieved when this solved the problem.[9]

In the meantime, the radar signal vanished from the screen. Soon afterward, Scott remembered, "the ship next to us was hit by something such as a missile or torpedo and was on fire." One report said that "apparently a Japanese aircraft . . . struck the ship below the bridge but above the water line," but Scott still thinks that "the stationary image on our radar . . . [had] something to do with the damage. . . . That ship was withdrawn and headed back to Eniwetok."[10]

After dawn broke on April 1, and an hour or two before the assault troops left the *Hendry*, a Japanese plane buzzed the detachment of American ships. Assuming it was an observation plane, the detachment commanders issued orders to antiaircraft gunners to shoot *toward* the enemy plane but *not* to shoot it down, because they wanted the plane to report the location of the force. Many ships opened fire on the plane, but it flew past unharmed.[11]

The amphibious feint on Minatoga Beach commenced within an hour or two after the Japanese plane passed. The *Hendry* carried two sizes of assault craft—two or three LCMs (Landing Craft, Mechanized) for heavy equipment and a dozen smaller LCVPs (Landing Craft, Vehicles, Personnel) for troops.[12]

Strict procedures governed the disembarkation process on the *Hendry*. Troops used both sides of the ship, climbing down heavy rope cargo nets at one of the three or four loading stations on each side of the *Hendry* and dropping down into the small assault craft bobbing on the surface below. They had to move quickly because the ship was particularly vulnerable at this time, but the cargo nets could not be overloaded with too many troops and the heavy packs they carried. If the *Hendry*'s captain was not satisfied with the speed of the disembarkation process, Scott recalled, he would start yelling at the junior naval officers in charge of the loading stations to "get them moving!"[13]

The fake assault on Okinawa commenced. The assault troops climbed down the *Hendry*'s cargo nets and into the small landing craft below.[14] The loaded assault boats then raced to within one hundred yards of a beach near the southern end of the island. Scott, who watched the action from the *Hendry*'s deck, described the north-south-running beach as quite flat with no cliffs.[15] He saw no fire coming from the Japanese defenders. As the small landing craft neared the shore, U.S. Navy destroyers sailing parallel to the beach cut in front of them and laid down clouds of smoke, hiding both the approaching landing craft and the U.S. fleet in the distance from the view of the defenders. The small American assault craft then suddenly made U-turns and returned to their mother ships still laden with troops, who scrambled back up the cargo nets.[16] The feint was finished.

Marine Corps historian Joseph H. Alexander concluded that the decoy operations achieved their purpose. Japanese commander "Ushijima retained frontline infantry and artillery units in the Minatoga area for weeks thereafter as a contingency against the secondary landing he fully anticipated." If anything, noted Alexander, "the deception proved too successful" because the Japanese "vectored a flight of kamikazes against the small force." The suicide bombers struck two ships, causing more casualties to the 2nd Marine Division on April 1 than were suffered by any of the divisions that actually landed on Okinawa that day.[17]

The real landing took place on the other side of the island on Easter Sunday, April 1, at 8:30 a.m. Shaped like a twisting serpent, Okinawa measures about 60 miles from north to south and 2–18 miles from east to west. The landing beaches—situated in a narrow section of the island roughly one-third of the way up from the southern tip—stretched for 15,000 yards from the north to the south.[18] Four U.S. divisions landed side by side on that April's Fools Day. The 96th Division had been assigned the southernmost landing beaches at the far right flank. Next came the Army's 7th Division, the 1st Marine Division, and finally the 6th Marine Division. Captain Bollinger, Lieutenant Sheahan, Lieutenant Burrill, Lieutenant Walton, and Private Moynihan were among the troops preparing to land.

Walton woke at 4 a.m. on April 1 for a preinvasion steak breakfast. After a brief church service, the soldiers spoke a few words of farewell to colleagues assigned to other landing craft. "This is it," they murmured to one another.[19] It was still dark as Walton and thousands of other Americans, loaded with gear and weaponry, scrambled down cargo nets and into the small landing craft that would speed them away from the transports.[20]

Lieutenant Sheahan landed very early in the operation. He went ashore with a team consisting of an artillery forward observer (himself), a naval gunfire officer, and an air liaison officer. Decades later he still recalled the emotional charge generated by the assault. "I was in the first . . . boat to land—we ran across the beach yelling—like all troops have done . . . down through the centuries."[21] An eerie silence met the Americans' shouts. The beach was empty save for the invaders. "There was nobody firing at us—there was nothing. We wondered where everybody was. We thought that we were going to be murdered . . . like they were on the other islands."[22]

Radio operator Charles Moynihan assaulted the beach between the first and second waves on board the infantry battalion commander's boat—called the "00" boat. Lt. Col. Russell Graybill, commanding the 2nd Battalion of the 381st Infantry, 96th Division, could have landed any time between the first and seventh waves and had chosen to come in early. A seawall ran parallel to the coastline at the point where Moynihan landed. Emerging from his landing craft, Moynihan needed help from other members of the landing party to carry his heavy Model 110 artillery radio over the barrier.

Lt. Donald Burrill landed in the second wave that morning, slogging ashore with B Company of the 381st Infantry, 96th Division. Their landing was easy. They worked their way inland, checking potential hiding spots and heavy cover for defenders, and dug in for a night position about halfway across the island. About that time Burrill received a message from A Battery saying that they were ready to fire. When Burrill asked them if they were located in the position assigned to them by the plan, the battery—perhaps still dazed by the quick pace of the landing operations—replied, "We hope so." He was able to adjust fire properly, and no attacks or other trouble disturbed that first night.[23]

Capt. Willard Bollinger and his infantry company came ashore shorthanded, having left the two rifle-less soldiers on the troop transport as a reserve. As it turned out, the rifles were not needed. During the first six days of the invasion, F Company of the 2nd Battalion suffered no casualties that Bollinger could remember.[24] The remainder of the campaign, however, would be an altogether different story.

SSgt. Karel Knutson did not participate in the early landings. His LST remained about ten miles offshore. His howitzer crew—loaded on board a DUKW (amphibious truck)—did not leave the ship until about 2 p.m. Knutson smuggled a five-gallon can of coffee onto the DUKW, even though it was not part of the authorized equipment. Knutson loved his java. His section enjoyed high-quality coffee throughout the campaign. Although Knutson normally commanded a gun crew of six, the limited space available on the DUKW dictated that he make the initial landing with only five. He selected an artilleryman named "Tom White" as one of the crew. Knutson considered "White" a good cannoneer and soldier. He did not know that some lower-ranking soldiers had secretly been producing alcohol on the LST and that "White" had imbibed heavily. He was drunk when he got into the DUKW.[25]

The invasion day went smoothly for Knutson's group, whose DUKW took less than two hours to cover the ten miles to the shore. Knutson could not recall the name of the beach where he landed, but he remembered a Japanese naval gun on the bluff to the right of his disembarkation site. His men met little resistance save for an occasional Japanese shell.[26] Knutson's experience was typical. Joseph Alexander noted that the assault troops "found no mines along the beaches, discovered the main bridge over the Bishi River still intact and—wonder of wonders—both airfields relatively undefended."[27] Knutson nevertheless heard rumors that the Japanese planned to let all the Americans land and then keep them fighting for a year.

B Battery went inland about two miles and set up its guns. The night passed without incident for Knutson's unit. During the night, however, they heard firing from A Battery, which was positioned to their right. The next morning Sergeant Knutson and some others got permission to investigate. An astounding sight greeted him. During the night the Japanese had attacked the A Battery perimeter with female soldiers. Someone had spread the naked legs of what was left of one corpse atop a rock. Knutson could see for himself that the attackers had been women in military uniform. They had worn explosive "vests," which the defenders' bullets had ignited.[28]

Lieutenant Walton did not make it ashore on invasion day. Although they had risen early, Walton and the artillerymen in his craft had to undertake an additional journey before they landed on the beaches. They had sailed from the Philippines in a different transport vessel from the main body of the 361st Field Artillery Battalion and had expected to transfer to that unit's LST before the dawn landing. "Our coxswain got mixed up, went the wrong way, and took us to the wrong anchorage," Walton recalled, so they did not land at all on the first day of the battle. The navigational error provided Walton and his shipmates with a spectacular view of the full breadth of the assault landing because the errant coxswain "had to cross the whole division front to get to the other anchorage." Dawn was just breaking. "About halfway the waves of Infantry LCIs [Landing Craft Infantry] came from our right and passed to our left as we weaved between them."[29]

By the time Walton and his men arrived at the correct LST, the artillery battery they were to accompany had already departed for the shore. The captain commanding the battery had gone with them, transporting the howitzers in DUKWs. Walton received instructions to remain on board as the invasion took place. "My orders were to stay on the LST until the beach was clear, it could land, and to make sure our equipment came off." Walton and his men faced a further delay when the captain of the LST "did not want to land because of a coral reef. (The captain was correct, some days later the landing was attempted, the ship grounded, and was never freed.)"[30]

Walton attempted to communicate his situation to his superior officer by means of the amphibious trucks that were shuttling back and forth between the transport ship and the landing beaches. "DUKWs came back and forth to pick up ammunition and I would send messages to the Captain [Rollin Harlow]—battery commander—'what do you want me to do' and would get no answer or 'stand fast,'" recalled Walton.[31] The combination of the navigation error by his initial coxswain and the understandable refusal of the LST captain to attempt a landing across a treacherous reef kept Walton at sea the first day of the battle.

During the initial landing, artillery forward observation depended heavily on observation aircraft. Since no airfields had yet been captured and the aircraft carriers were busy handling naval aviation, the flimsy artillery spotting aircraft had to take off and land using a "Brodie device" on an LST that Walton described as "a suspended cable under which the planes landed by means of a hook. When they took off, they would rev up the engine, release some type of brake, go at full speed to the end of the cable, drop about three feet, flutter a few seconds, and fly away."[32]

The Japanese had long realized that they could not prevent an initial American landing, so they tried to wreak a horrible punishment on the vulnerable American invasion fleet lying offshore by vectoring in squadron after squadron of Kyushu-based aircraft. As night fell, Walton watched from his LST as Japanese aircraft assailed the American invasion fleet off the coast of Okinawa. Fireworks filled the sky like a giant Fourth of July show, he said. "Several ships were hit even though the Navy put up an ack-ack antiaircraft umbrella over the fleet. One of my men was injured by a piece of an ack-ack projectile and had to be treated by the ship's doctor." On April 2, Walton finally received instructions to land. "My battery commander [Harlow] requested me to come ashore the next morning and I didn't see sheets again for at least two months."[33]

A triumvirate of Japanese officers conducted the defense of Okinawa. The Japanese commander, Lt. Gen. Mitsuru Ushijima, was tall, dignified, and serene, the personification of the classic samurai warrior. He had once commanded Japan's military academy. Now he led the Japanese 32nd Army. His chief of staff, Lt. Gen. Isamu Cho, was fiery, demanding, and volatile.[34] Unlike Ushijima, Cho was closely associated with the ruthless right-wing militarists who dominated Japan in 1945.[35] Together the two men made a good team; each complemented the other's strengths. A brilliant planner, Col. Hiromichi Yahara, served as the operations officer and senior staff member of the 32nd Army. Quiet, deliberate, intellectual, and aloof, Yahara got along well with General Ushijima but frequently clashed with General Cho.[36] Only Yahara survived the battle, and his account of the Japanese strategy has long dominated historical discussions of it.

The lack of Japanese opposition to the American landings was no miscalculation. In late 1944 the Imperial General Headquarters in Tokyo had moved the best Japanese division on Okinawa, the experienced 9th Division, to the island of Formosa (Taipei).[37] Lt. Oliver Thompson later remarked, "Boy, can you imagine the carnage on Okinawa had they remained?"[38] At the time, however, the move was based on a reasonable estimate of American intentions. Indeed, the Americans had once planned an invasion of Formosa, but Admiral Nimitz obtained permission to bypass that island in favor of an immediate invasion of Okinawa.[39]

General Ushijima and his staff thus prepared Okinawa's defenses with one less division on hand. Accordingly, he chose to defend the landing beaches only with lightly armed troops called *boetai*—many of whom were native Okinawans (as opposed to most of the other troops, who came from elsewhere in Japan).[40] Despite having been stripped of his finest division, Ushijima had approximately 100,000 Japanese fighting men available to oppose the invaders (67,000 from the Japanese Imperial Army, 9,000 from the Japanese Imperial Navy, and 24,000 armed Okinawans).[41] Imperial General Headquarters in Tokyo hoped to save Okinawa by destroying the American fleet and dispatched both conventional and suicide planes to sink it. The more realistic defenders on the island recognized that the conquest could only be delayed, not denied. Ushijima and his forces thus prepared to defend the island to the death.

American campaign strategy called for the immediate acquisition of airfields—specifically, those at Yontan and Kadena near the invasion beaches. Both were seized on the first day, and Okinawa rapidly became a sort of unsinkable aircraft carrier. Next, General Buckner ordered his troops to cut the island in half. This goal was facilitated by the selection of landing beaches near one of the island's narrowest sections. Once the island was split in two, Buckner planned to dispatch the Marines north and the Army south. Hence the northernmost landing beaches were allocated to the Marine divisions. While no one could be certain where the Japanese would mount their strongest defense, the north was expected to be an easier conquest than the south. If this proved correct (as it did), the fast-moving Marines might more easily be supplied by sea as they fought their way north. The Marines promptly subdued the northern section of the island. By May they were repositioned in south-central Okinawa helping the Army forces attempting to break through the Japanese defenses around Shuri Castle.

American strategists had several reasons to believe that the Japanese would mount their major defense in the southern half of Okinawa. It was the most densely populated section of the island. Naha—the largest city and the provincial capital—lay in the south. In particular, Japanese troops could be expected to vigorously defend Naha's fine harbor against American seizure. A series of east–west ridges stood between the Army troops in the center of the island and Naha. (The Americans would give these heights names like Kakazu, Hacksaw, Wana, and Shuri.) The hard-working defenders had burrowed into these strong points, making them unsinkable battleships bristling with weaponry. From these defenses the Japanese planned to deliver broadsides of deadly fire against the Americans.

General Buckner directed frontal assaults against this series of defenses. He concentrated massive firepower against them, sending flamethrowers to drive the Japanese back from cave entrances and demolition teams to seal the defenders inside. General Buckner's decision to breach the central Japanese defense line was controversial.[42] He might, for instance, have elected simply to seal off and isolate the southern third of the island. General MacArthur, viewing the situation from the Philippines, asserted in private that the Americans should have stopped when they came up against the well-entrenched defenses and starved out the Japanese defenders. "The Central Pacific command just sacrificed thousands of American soldiers because they insisted on driving the Japanese off the island," he said. "In three or four days after the landing the American forces had all the area they needed, which was the area they needed for airplane bases.

They should have had the troops go into a defensive position and just let the Japs come to them and kill them from a defensive position, which would have been much easier to do and would have cost less men."[43]

Of course, MacArthur's scheme would have resulted in a long delay before the Americans gained control of Okinawa's two major ports: Naha and Yonabaru, and their adjacent anchorages, which could assist the naval buildup preceding the invasion of Japan. Moreover, the Central Pacific campaign was dominated by naval planners and led by naval admirals. Thus the Tenth Army positioned itself to seize all of Okinawa, including the line of ridges that protected Shuri and the island's two major harbors.

With enormous consequences for the future, this strategy brought the Americans in contact with the entire island and its people. In May 1945 Lieutenant Walton took advantage of a pause in the battle to write a detailed description of the island's geography for his wife.

> The island is really separated into three parts, the northern, central, and southern part. . . . The southern part is the most densely populated sector. It contains the three large citys [*sic*] of which Naha is the biggest. Naha has about 65,000 people. . . . There are no real flat and level spots on the island, it's mostly little hills or foothills, just mounds and little valleys in between. . . . Practically all the island is under cultivation. The Japs always were energetic people and produced beautiful gardens. These are by no means different. They have all of the hills terraced with stone walls and well planted. . . . As I have told you before we have quite a variety of trees fir, pine, etc. . . . All [in] all it's kind of a temperate tropical climate. . . . The goats have a lot of small kids right now and the kids are really cute.[44]

The uneven nature of the Okinawa's terrain proved a boon for the Japanese defenders and a bane for the Americans.

CHAPTER 4

ASSAULT ON KAKAZU RIDGE

American forces cut across Okinawa from west to east within three days, isolating the island's northern and southern halves. Lt. Charles Sheahan confirmed that "there was a little scattered firing here and there. But we cut across the island very quickly. It was only about three miles [wide] where we landed." No one opposed them. "Instead of a second Iwo," Sheahan recalled, "just a few pop shots—a surprise."[1] Pfc. Charles Moynihan was also surprised by the absence of opposition as the Americans cut the island in half. After the first day, his infantry battalion had suffered only one or two casualties.[2] Opposition remained light for the next few days, leading some to designate this brief period as "honeymoon week."[3] William Filter, a twenty-year-old rifleman in Capt. Louis Reuter Jr.'s G Company, remembered being all the way across Okinawa by the second day after landing. Indeed, he heard rumors that the Japanese had evacuated the island. Soon those rumors would be confounded: furious fighting soon cut bloody swaths through the ranks of G Company. The combination of the casualties and Filter's stalwart conduct in the face of the terrible fighting would propel his rise through the ranks from private to staff sergeant. His job responsibilities increased accordingly from rifleman to squad leader to platoon leader.[4]

The mild resistance met during honeymoon week resulted in new orders for the artillerymen. The original plan had called for the artillery to face east toward the initial infantry effort to cut the island in half. B Battery of the 361st Field Artillery Battalion had placed its howitzers in firing position but was called back because it was not needed on the line. The staff directed the realignment of the artillery to face south toward Naha. Soon the U.S. Army infantry had also turned south to face Naha, with the 96th Division on the right and the 7th Division on the left.[5] Japanese resistance increased as the 96th Division moved farther south.

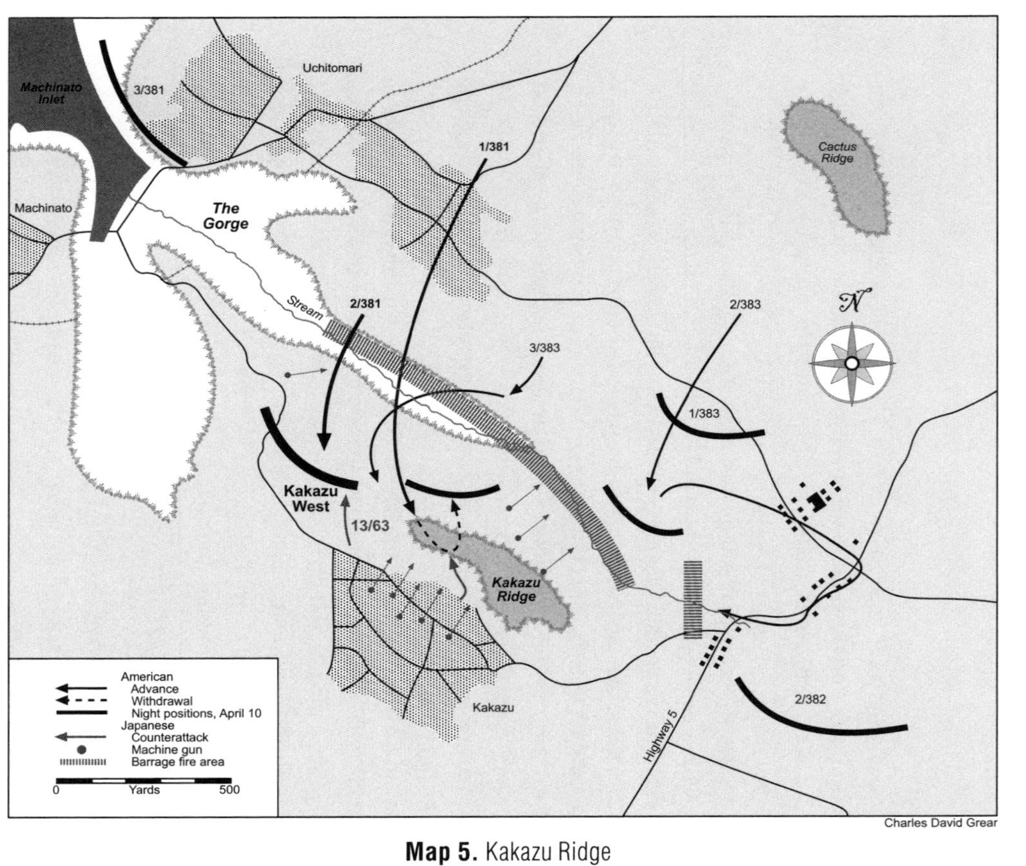

Map 5. Kakazu Ridge

Adaptation by Charles Grear based on map X in Roy E. Appleman, James M. Burns, Russell A. Gugeler, and John Stevens, *Okinawa: The Last Battle* (Washington, D.C.: Center of Military History, 1948, 1991).

As the Americans moved toward their objective, artillery came increasingly into play. A forward observer named Lieutenant Ward was killed by enemy rifle fire.[6] The problems with inaccurate artillery fire on Okinawa began very early in April too. Lieutenant Sheahan was directing howitzer fire in support of a tank unit when he himself became the subject of friendly fire. Indeed, without the protection of an armored vehicle, Sheahan might not have survived the errant shells. "They bounced on top of us. I would say maybe ten or twelve rounds. . . . I really wouldn't know if anybody got hit. . . . You are very much by yourself. You're there but you don't know what's going on in the bushes over there or over here."[7] Sheahan "gave them [the artillery battalion] hell back there" for the mistake.

The "short-rounds" incident demonstrates the complexities of the relationship between the forward observer and the maneuver units (infantry and armor) he accompanied. If Sheahan had not reacted strongly, the armor personnel might have assumed that his mistake had caused the friendly fire. Moreover, an artillery officer who was seen as too forgiving and understanding about errors from his own branch might soon lose the confidence of his infantry and armor counterparts.[8]

In the meantime, Lieutenant Walton had finally joined his artillery battery, although he was not immediately assigned to a forward observation position. During the period from April 3 to April 8 he served as the temporary executive officer (second in command) for his battery and was responsible for firing the battery's four howitzers during daylight hours. At night, the battery commander replaced him. Night fire usually consisted of prearranged defensive fire conducted on an hourly basis. Walton occupied his spare time writing letters. "In between firing missions," he remembered, "I would try to write a letter to [Carolyn, his wife] each day, which was not easy, as all I could say was 'somewhere in the Ryukyus' for a location." Censorship restrictions also hampered his ability to describe his current situation. "I could not say a thing about the battle that I was totally committed to because of security reasons."[9]

On April 9, Walton and his observation team "went forward with the infantry for the first time." He got "into position, dug my slit trenches, and spread my poncho over the top. Good thing, as it rained that night." In some theaters of World War II, forward observation teams made frequent use of team jeeps that remained parked not far behind the line of battle. The steep terrain on Okinawa, however, frequently dictated a different practice. During the early morning hours of April 10, Walton and his team left their jeep behind for good and moved out on foot.[10]

Lieutenant Sheahan's unit headed south, toward a series of east–west escarpments that Sheahan described as "perfect defensive positions" because they ran across the entire island, blocking the path of the Americans. An American intelligence officer on the scene described Kakazu as "one of the most formidable terrain barriers ever encountered in battle."[11] Kakazu Ridge and Kakazu West, joined by a saddle, dominated the first escarpment Walton encountered. Kakazu's steep northern slope was a formidable climb, made even more difficult because one first had to cross a gorge with a stream running through it. The Japanese rained down fire from fortifications near the summit.

Kakazu Ridge came first. That ridgeline still loomed large in Sheahan's mind more than half a century later. "Ten days after landing we ran into Kakazu. Of all the hills, that will always be number 1." Kakazu Ridge, he said, "took a lot of our time."[12] The Japanese defenders chose well in making this ridge their initial defense line. Seeking to thwart the Americans' advance toward the ancient Okinawan capital city of Shuri, just a few miles northeast of the modern capital of Naha, they had created an interlocking defense system in which the weapons on one ridge protected the approaches to the adjacent ridges. The Japanese army headquarters lay deep underneath Shuri Castle.

The honeymoon ended on April 9. The 383rd Infantry Regiment, which like the 381st Regiment belonged to the 96th Division, led the attack on Kakazu. The troops had to cross the gorge under concentrated Japanese mortar fire. Lt. Willard "Hoss" Mitchell, a company commander in the 383rd Regiment, was awarded the Distinguished Service Cross—the nation's second-highest military award—for his actions on Kakazu West. His isolated company turned back four Japanese counterattacks. The Japanese, however, "had more men, more weapons, more ammunition, better ground," and Mitchell's L Company withdrew under the cover of smoke and artillery, leaving Kakazu West in the hands of the Japanese once again. The 96th Division history reports that only three of the eighty-nine men in the company returned unscathed.[13]

Pfc. Edward Moskala, also of the 383rd, received the Medal of Honor for his valor on Kakazu Ridge. He spearheaded the assault up the escarpment, then covered his company's retreat when it was driven off the ridge. Although he had reached safety, he returned across the gorge in an attempt to rescue wounded buddies. His heroism that day cost him his life.[14] The American assault of April 9 had been repulsed. Some of the infantrymen from the 383rd came back through the lines of the 381st Regiment as they retreated off the hill.[15]

The failure on April 9 did not end the effort to take Kakazu. The Americans renewed the assault the next day with a different regiment—the 381st. The assignment of seizing Kakazu West fell to the 2nd Battalion.[16] Artillery came immediately into play. Private Moynihan, attached to the 381st, had heard a few details about the previous assault. He understood that as many as five hundred Americans had fallen. Moynihan watched as artillery softened up the defenders in preparation for the morning thrust, laying down a heavy barrage. Moynihan vividly recalled the Japanese return fire. "Well, they used the 150 [150-mm artillery] which was mostly what we encountered and they had . . . a 'buzz bomb' but they had to be within a thousand feet to set that off. You could hear that coming and . . . you could watch the whole projectile go through the air."[17] Moynihan was in his usual post as radio operator for an artillery liaison officer when a Japanese 150-mm shell landed near their observation position.

> My captain—the one I had before Sheahan . . . we were sitting in a shell hole. . . . I had my radio up in front of me. And a 150 came and hit four or five feet away—or maybe a little more—and exploded and it knocked all the tubes out of my radio . . . and the captain, . . . he says, "I've been hit, I've been hit!"
>
> And so I asked him "let me see," you know, and what actually happened was that a piece of shrapnel had hit his dog tag and had skinned him. That's all that happened to him. But he thought he had been hit in the chest. He had been hit in the chest, but fortunately the dog tag [deflected it.] It bent the dog tag, I'll say that for it.[18]

After the close call from the Japanese artillery, Moynihan returned to the artillery battery for a new radio. The U.S. artillery barrage was still under way when he returned. "We laid down . . . about two hours heavy artillery prep, with the help of the Navy also. Then that all stopped and then because we start the rolling barrage and then we go underneath the rolling barrage. And the cannoneers raise the . . . base of the gun which gives you a longer range."[19]

The second day's assault on Kakazu Ridge began at 7 a.m.[20] The first man into the gorge at the base of Kakazu was killed, but one platoon managed to cross the stream and gain some protection from overhanging rocks on the south side of the gorge. Another platoon "made its way through poisonous mortar fire to join them there."[21] The 96th Division history credits five men from Capt. Willard Bollinger's F Company with destroying the strongest defensive position on the slope and allowing the company to reach the crest. E Company soon joined them. The battalion commander ordered up Captain Reuter's G Company to stabilize the situation. Attempts to seize "a strategic finger" that "ran west toward Kazaku" failed in the face of a strong Japanese counterattack.[22]

Some Americans temporarily escaped the full vigor of the Japanese resistance. Rifleman William Filter, advancing with G Company, faced only light opposition getting up Kakazu during the morning attack. Although there was scattered mortar fire, even the gorge was not fiercely defended at the specific moment he darted through. Filter perceived this as part of a Japanese strategy of pulling the Americans up onto the hilltop, where they would be exposed to Japanese fire and infantry attack.[23]

The second day's assault exposed Private Moynihan to the full weight of his responsibilities as an observer team member. When the infantry attack began, Private Moynihan accompanied his officer forward. His first problem consisted of transporting his heavy and unwieldy radio to his prescribed position. He had to cross an open field just to get to the base of Kakazu Ridge. American guns were firing a rolling barrage over the heads of the U.S. infantry as they advanced, and the shells were falling directly in front of Moynihan and his captain. The American artillerymen timed their fire to avoid hitting their own troops and keep the enemy pinned down, but the shells falling in close proximity could be an intimidating experience for the infantry. Moynihan and the battalion liaison officer ran right under the barrage. "And that one," recalled Moynihan many years later, "was a hairy one."[24]

Hauling the heavy radio proved too much for both Moynihan and the pack board on which he carried the radio. Because their position was so exposed, they would run a few steps and then drop to the ground. Moynihan explained, "We would count and then . . . dive." It was a painful process. Each dive slammed the pack board into Moynihan's shoulders. When the straps holding on the pack board finally broke under the stress, he salvaged new equipment from a dead GI. "I was out there with no straps and so I asked my captain to run over to a fellow who was killed who had a pack board on him."

With his pack board repaired, Moynihan and his captain continued their advance. They crossed a small knoll before reaching the stream at the foot of the ridge. "I had to be helped to get up there because of the weight," he recalled. "I had to have a couple of those infantry fellows help me because everything was so slick and muddy." After crossing the stream they faced a steep climb up the ridge. Near the summit, they found the colonel commanding the infantry battalion from a command post in a cemetery positioned in the draw between Kakazu and Kakazu West. The colonel's party was partially shielded from Japanese fire by one of the many large hollow, turtle-shaped tombs that dotted the island. The infantry was stymied forward of the command post. The front line ran along the top of the ridge (about where the large blue viewing tower on Kakazu Ridge is located today). The Japanese front lines and the U.S. front lines "were just only a few feet apart," Moynihan recalled.[25]

Lieutenant Sheahan, who also participated in the second assault, recalled that Kakazu was "very rugged country" with a "very sharp valley" in front. "It was a very deep gorge. . . . That's why it was so tough." Sheahan did not receive mortar fire while crossing the gorge, but he saw American "bodies and . . . rifles all over the place." Once across the gorge, like Moyhihan, he still had to climb the ridge proper with a fifty-pound radio on his back. "I couldn't adjust artillery because I couldn't see anything," he remembered. "But then I backed off and came down again. And we went into one of these . . . tombs . . . [when] a lot of Japanese artillery came over." Sheahan also recalled one of the peculiarities of the situation: "We used to take the [Okinawan] bodies out [of the tomb's interior]. They had them in barrels of fluid."[26]

By 9:30 a.m. the 2nd Battalion had reached the top of Kakazu West but was barely hanging on.[27] Brig. Gen. Claudius M. Easley, the Deadeyes' assistant division commander and coordinator of the assault, sent reinforcements. At 1:45 p.m. he dispatched the 1st Battalion of the 381st into the fray. The unit headed down toward the gorge at the foot of Kakazu but quickly ran into trouble. The division history notes that "the Japs counter-attacked with devastating results" while "at the same time the enemy was raising havoc with the 1st Battalion, 381st, as they attempted to move onto Kakazu. . . . [T]he battalion command group, including Colonel Cassidy, were trapped in the gorge by an impenetrable concentration of mortar and machine gun fire."[28]

A third observer team pressed forward at this crucial time. Lieutenant Walton and his men were in the gorge, "moving up through a stream bed when Japanese 60-mm mortars zeroed in on us. I was wounded along with two of my men," he wrote decades later. Walton's wound—some slivers of mortar metal in his back—"was quickly determined to be minor." Two of his team members had more serious injuries and were bandaged and sent to the rear. Mortar fire peppered Walton, the remnants of his observation team, and the infantry reinforcements trapped in the gorge. As night approached, an infantry captain encouraged them to rush across the stream, reminding them that they needed to get up the ridge in time to dig in for the night. Walton's three-man observation team moved out with two companies of infantry and made it up the ridge with "a few more casualties on the way."[29]

Once they got within roughly one hundred feet of the summit, Walton's team found a spot to spend the night. Leaving one man at that location, Walton and his radio operator climbed up another hundred feet to an observation point near the top of the ridge to assess the most likely avenues of approach for a Japanese attack. He then zeroed in artillery fire on those locations. The Americans made a practice of sporadically firing on these sites during the night, both to interdict concentrations of Japanese troops and to harass Japanese supply movements.

Night had fallen by the time Walton had arranged the necessary coordinates with the FDC back at the artillery battalion. Walton and his radioman were at that point in deadly peril from their own troops because they had advanced beyond the American line of battle. On the front lines, it was typically the Japanese who moved at night. American troops usually remained in their foxholes after dusk with orders to shoot anything that moved. If Walton and his radioman remained where they were, however, they would be on the very front lines without even the standard infantry weapons to protect themselves and would spend a chilly night potentially exposed to Japanese artillery fire. They would also risk being caught in the crossfire between attacking Japanese and defending Americans. The lieutenant and the radioman decided to move down the hill toward the American front lines, attempting to communicate the fact that they were forward observers, not Japanese attackers. "All we could do was yell 'FO! FO!' and move down the ridge." To Walton's great relief, no one shot them as they made their way back. He was lucky. A forward observer with the 1st Marine Division on Okinawa reported that a panicked Marine killed five of his fellow Marines in a friendly fire incident before he himself was shot.[30]

Meanwhile, down in the Japanese cemetery, Private Moynihan faced abnormally onerous communication duties. His battalion-level artillery liaison team had to relay radio messages from the company-level forward observer teams. "Some of the radios . . . from the forward observers would not reach the fire direction center from the top of Kakazu Ridge because there was some interference in there that drowned them out, and . . . I had to relay their fire missions for them," Moynihan recalled. Moynihan attributed some of the problem to the rainy weather. After the first afternoon on Kakazu Ridge, however, the problem disappeared and the other radios were able to communicate directly with the FDC.[31]

The battle for Kakazu Ridge lasted nearly two weeks. The effort to hold the summit demonstrates the tactical and practical limitations of artillery on the World War II battlefield. The 96th Division history sums up April 10 with the words: "In another day of grim and bloody combat, only the 2d Battalion of the 381st had made an appreciable inroad in the Japanese positions."[32] The Japanese slowed the attackers by employing "reverse slope" tactics throughout the Okinawa campaign. That is, they fortified not the slope facing the American assault (where the Japanese forces would be subject to direct fire from American forces) but rather the back side of the slope where the defenders were protected from all but indirect fire.

Thomas Huber and others have noted the resemblance between the Okinawa campaign and the trench warfare of World War I.[33] They have also pointed out the differences.

> Because of the undulating terrain, [Japanese army] units were able to build their forts into the hillsides while still giving a view on the world that was above ground level, and that seemed to and did dominate terrain. This, plus the fact that the caves were almost completely safe from bombardment, seems to have spared [Japanese] soldiers some of the dismay that World War I soldiers experienced living day in and day out in the trenches. The Japanese soldiers did not suffer as much as the World War I soldiers from the "underground neuroses" described by Eric Leed.[34]

The Japanese army remained in fortified positions atop and on the reverse side of Kakazu even as the Americans held the front (north slope), and counterattacked, usually at night.

For infantry commanders such as Captain Bollinger, Okinawa became a battle of "one hill after another." The Japanese were skilled in making use of the reverse slope defense and in fortifying numerous small knolls. The Japanese were so well dug in, Bollinger later quipped, that they might still be there except for their wasteful banzai charges. Bollinger considered a combination of artillery and air attacks with napalm the best method to reduce the threat they posed.[35]

Artillery proved less effective when the fighting was at very close quarters (as it was on the crest of Kakazu). Lieutenant Sheahan noted that the Japanese on Kakazu were "well dug in and they had a lot of machine guns" and attributed the length of the struggle to the strength of the Japanese defenses and their determination to resist: "It was a very tough nut to crack." The Japanese fended off more than one American outfit. "It wasn't only the 96th involved in Kakazu," Sheahan recalled. "It was a long ridge too, it wasn't just a hill." At first the target for U.S. artillery "was just the top of the hill," but when the Deadeyes reached the plateau on the summit, "we couldn't fire anything" for fear of hitting friendly forces.

Sheahan noted particular problems of enemy proximity complicated by geography. The fierce fighting on the summit meant that no safe vantage point was available there for the artillerymen. Sheahan could not even get up there for "a couple of days," he recalled. Once on top, he continued to experience difficulty directing artillery because of the short distance between American lines and Japanese lines, and because he could not see down the reverse slope from his position. "It was up to the infantry," he said.[36]

Sheahan remembered a small draw in the hill near the top of Kakazu Ridge where a trail ran between the Japanese lines and the American lines. "That was a hot spot. . . . The 1st and 2nd Battalions were operating in that area. . . . They were overlapping each other." The front line was just over the summit of the ridge and down onto the southern side. Both sides fought ferociously over this relatively small plot of ground.[37] On the night of April 12, SSgt. Beauford (Snuffy) Anderson, a mortar man from the 1st Battalion of the 381st, repulsed the attacking Japanese first with his carbine and then by throwing mortar rounds, which are normally fired through a mortar tube. He received a Medal of Honor for his heroism.[38]

The ferocity of the Japanese counterattacks tested the Americans' resolve. Artillery radioman Moynihan and the infantrymen positioned in front of him withstood two banzai attacks on Kakazu, furious human-wave assaults by Japanese light infantrymen who rushed forward shouting, "Banzai!" in honor of their emperor. Private Moynihan was at the 2nd Battalion command post, located about 100 to 150 feet behind the infantry companies, but that did not protect him from Japanese attempts to retake the crest of the ridge and penetrate the draw. The Japanese troops typically did not expect to survive a banzai assault and pressed forward despite mortal wounds. Decades later, Moynihan retained vivid recollections of unusual methods adopted by the Japanese to keep going even when wounded: "When they came over . . . they [already] had tourniquets on their legs, by their groin, and they had tourniquets on their arms."[39]

The infantry in front of Moynihan made a valiant effort to stop the attackers from penetrating the American lines. "The machine gun that I was behind—of the infantry—we counted seventy-eight stacked up in front," he recalled. The American defenses proved insufficient to stop the hard-charging Japanese. "They ran right through us—some of them," Moynihan remembered, all the way past the machine-gun position on the crest of the ridge and through all of the infantrymen positioned between the front lines and Moynihan's position behind the lines.

Japanese soldiers who survived the assault and succeeded in recapturing Kakazu Ridge clearly planned to stay. Each carried a three-day supply of food. Their counterattack did not succeed, though, and the Japanese attackers paid a terrible price. "They were stacked up like a bunch of worms," Moynihan recalled.[40]

Moynihan remembered yet another counterattack in which a Japanese officer ran in front of and parallel to the U.S. lines. "On that last banzai at six o'clock an officer got all dressed up . . . and ran straight across in front of us. . . . I didn't realize it but he was committing suicide," Moynihan said.[41]

Filter, by now a sergeant and an infantry squad leader, was with the unit that held the western end of Kakazu West—the far right of the American line on the ridge. He recalled that the Japanese knew exactly which unit they were attacking, shouting, "G Company Banzai" before they charged. The attackers met with less success here than at Moynihan's position. All of the charging Japanese soldiers were mowed down before they got within twenty-five yards of Filter's position.[42]

The struggle for the summit of Kakazu Ridge left indelible memories in the minds of the men who fought there. Moynihan, who served the entire Okinawa campaign, called it "the worst combat that I was in," noting the number of men killed. "There was hand-to-hand fighting there," Moynihan said, although he did not personally witness it. The U.S. troops threw hand grenades, and the Japanese picked them up and threw them right back, he remembered. A cave located in the saddle between Kakazu and Kakazu West was pressed into service as a makeshift hospital and to store surplus weapons from the combat-incapable American casualties.[43]

The Americans faced shortages of everything from basic supplies to artillery shells after the Japanese disrupted efforts to get supplies and men up Kakazu Ridge. A "human pack train" brought supplies up to the 2nd Battalion, suffering fifteen casualties on the way.[44] Private Moynihan found his unit cut off from even medical evacuation services.

> The Japanese had us encircled. A fellow that was in my section got hit. We couldn't get him a stretcher because they had us completely surrounded. . . . I was in my foxhole and a two-star general got in with me. . . . I asked him what he was doing and he said, "Well, I just came from Europe." [The general was on an inspection tour and was en route to Washington, D.C.] Of course, he couldn't get out. . . . He said, ". . . I've never seen anything like this in Europe." And he says, "I don't know how you fellows even hold it here." And he says, "I've never seen anything as bad as this!". . . He was in my foxhole for a couple hours, but as soon as we were able to get a stretcher through he took off.[45]

The fighting on Kakazu created supply problems in the rear areas as well as on the front lines. American forces were using up artillery shells at a substantially greater rate than planned.[46] Worse yet, Japanese suicide bombers sank two ammunition ships (*Hobbs Victory* and *Logan Victory*) on April 6, resulting in a shortage of both 105-mm and 155-mm artillery shells.[47] General Buckner had anticipated the massive use of firepower on the island, but even he had underestimated the demands that would be made on artillery during the battle.

Moynihan vividly recalled the shortage of ammunition for the mortars and artillery. "As I remember we only had a little over one hundred rounds per battery and they were flying in mortar rounds by air from the Philippines."[48] Air transport of ammunition was unusual at that time. Heavy material usually came by boat.

On April 10 and 11, Maj. Leon Addy, the supply officer (S-4) for the 381st Regiment, reported a critical shortage of 105-mm field artillery ammunition and 81-mm mortar ammunition.[49] On one occasion Moynihan called for artillery support, but the artillery battalion had already used its ration for the day. The colonel commanding that unit tried to get permission from higher up to exceed the day's ration, but his communications with his superiors were cut. Risking reprimand, the artillery colonel went ahead and exceeded his allotment. Even so, the batteries soon ran out of ammunition again. Moynihan turned to the offshore fleet. "I begged the Navy . . . to fire on us. . . . I wanted them to fire right on top of us . . . but he wouldn't come in more than six hundred yards. And so the Navy really didn't do us much good there on Kakazu Ridge."[50]

Artillery did have successes on Kakazu. Regimental records indicate that on April 14, artillery support stopped the Japanese. At 7:20 p.m. Japanese forces began forming up in front of the 2nd Battalion's position after laying down a heavy artillery and mortar barrage. Within ten minutes, however, artillery stopped the counterattack. Artillery and mortar fire stopped a second counterattack at 11:30 p.m.[51] These actions were perhaps the artillery's single most important contribution in support of the 2nd Battalion on Okinawa. Perhaps the most important overall contribution of American artillery in the campaign was simply the day-to-day use of fire that destroyed enemy positions or at least made the Japanese take cover so that American offensive operations were possible.

With carnage all around them, the death of one American took the soldiers by surprise. While the nation collectively mourned the death of President Franklin Roosevelt on April 12, the battle raging on Kakazu left the Deadeyes no time for reflection.[52] Later, Captain Bollinger found a Japanese propaganda leaflet written in English that falsely reported that President Roosevelt had committed suicide. A Japanese Betty bomber had dropped it on a daring low-level night flight that managed to slip past the American air cover, which generally maintained air superiority over the skies of Okinawa. Propaganda failed to shake the Americans' foothold on the summit of Kakazu, but the Americans likewise failed to eject the Japanese defenders from the southern half of the ridge. Stalemate.

CHAPTER 5

DAILY LIFE

The artillery observers' experience on Okinawa cannot be fully understood merely by narrating the course of the artillery battle. Their roles and activities were far more complicated than merely calling in fire missions. The observers had to interact with two different combat branches—the infantry and the artillery—on a daily basis. They had to work closely with both officers and enlisted men, interacting with both the low-ranking personnel tasked with fighting the battle and the higher-ranking personnel charged with conducting it. Comprehending the organizational structure within which the observers functioned is also important to understanding their experience. Forward observers were a new part of battle tactics in World War II, and the structuring of artillery operations developed on a somewhat ad hoc basis to meet wartime demands.[1]

The hierarchical structure did not always function as planned. Boundaries between ranks and job functions sometimes blurred. The artillerymen nevertheless proved flexible enough to adjust their practical needs to the situation at hand. Although strict military formality and courtesies were not always followed, the arrangements proved adequate to meet the severe challenges presented by the Okinawa campaign.

Four aspects of artillery life on Okinawa merit close examination: the formal hierarchical structure of frontline artillery support; the blurring of the distinction between artillery liaison teams and forward observation teams; the weakening of the formal relationship between artillery officers and men on the front lines; and the frontline artilleryman's daily life, including the distinction between his roles in daylight and after dark.

Field artillery batteries were the "home" units for the observers and their team members. A "firing battery" typically consisted of four sections, one for each of the battery's four howitzers. At full strength, such a unit consisted of roughly one hundred soldiers.[2] Typically, a captain led a battery with a senior lieutenant as second in command (executive officer) plus additional junior officers. The field artillery observation teams were typically formed around these junior officers (see chart A). It was practical to have low-ranking officers in such exposed positions because they were the most numerous and expendable officers. In addition, low-ranking officers were usually younger and fitter than senior officers.

In addition to their home battery, the observers also interacted directly with the next-higher echelon—the field artillery battalion.[3] A light (105-mm howitzer) field artillery battalion in World War II typically comprised roughly five hundred men and consisted of three firing batteries. In addition there was a Service Battery plus a Headquarters and Headquarters Battery, neither of which had howitzers. The individual firing batteries of the 361st, for example, did not have their own fire direction centers. Only one FDC existed for the entire artillery battalion. Pfc. Charles Moynihan explained its operation. "We called to the fire direction center, which was the headquarters. They plotted a vertical and horizontal. When we called down, they would give the coordinates. . . . They would plot it. . . . [The forward observer] would tell them how many rounds he wanted, and then it would go down to the guns. They would set the guns—do the firing."[4]

In peacetime conditions, off-duty officers and men had been strictly segregated. Officers ate, slept, socialized, and relaxed in different locations than enlisted men did. On larger bases, the noncommissioned officers were often segregated from the lower-ranking enlisted men. Soldiers' friends tended to be men of their own rank. Having friends one rank above or one rank below was not unusual, particularly with wartime promotions, but friendships rarely spanned wide variations in rank. Soldiers of all ranks were expected to be familiar with military customs and courtesies. Enlisted men were expected to salute all officers (although not other enlisted men). Officers were expected to return the salute. Officers were expected to salute higher-ranking officers. In peacetime and in rear areas, all soldiers were expected to make a prominent display of their insignia of rank. Glittering rank insignia immediately identified the senior person(s) in any group. Combat in the Pacific,

CHART A
FIRING BATTERY STRUCTURE (SIMPLIFIED)

Battery Commander (captain)
Executive Officer (usually a senior lieutenant)
Reconnaissance Officer (often used as a forward observer; usually a lieutenant)
Forward Observer (lieutenant)
Assistant Executive Officer (often used as a forward observer; typically a second lieutenant)

First Sergeant (the senior enlisted man in the battery)
Mess Sergeant
Supply Sergeant
Motor Sergeant
4 Section Chiefs (sergeants who each commanded one howitzer gun crew)
Other Enlisted Personnel

however, required forward observers and other artillerymen to adopt different behaviors.[5]

The observers did not work by themselves; they functioned as one component of a team. The composition of such teams evolved as the war progressed. By the time of the Okinawa campaign, the U.S. Army had more than two years of combat experience with the new mobile forward observer system. Some of the kinks in the system had been worked out in places like North Africa and Guadalcanal during late 1942 and early 1943.[6]

An artillery forward observer team on Okinawa typically consisted of four or five men: the observer, a head of section, a wire telephone operator, and one or two radio operators (see chart B). Two of the team members were "permanent" in the sense that they generally accompanied one another. The highest-ranking of these was the observer; the other was the head of section.[7] One of the enlisted men also served as a jeep driver if the team was issued a jeep. If the team had no jeep, the radio operator had to carry his equipment—a particularly difficult job. The radio was so heavy that it was broken into two parts to be carried on pack boards.

CHART B
FORWARD OBSERVATION TEAM STRUCTURE

Observer (usually a lieutenant)
Head of Section (ideally a sergeant or corporal)
Wire Telephone Operator
Radio Operator (1 or 2)

If either half was lost or damaged, the radio was useless.[8] A heavy spare battery often had to be carried, too.

Combat conditions on Okinawa created an enormous demand for forward observers. Infantry commanders constantly pressured artillery commanders to provide more of them.[9] While the battle still raged, Brig. Gen. Robert G. Gard, the commander of the 96th Division's artillery, told artillery colleagues back in the United States: "You can't get too many people trained to adjust fire. . . . Our infantry wants an FO party with every rifle company—even those in reserve. We just don't have enough men to do it. Every battery ought to have at least 15 officers and non-coms trained to handle a forward observer assignment. That means other enlisted men must be trained to do the wire and radio work. You never have enough."[10]

The firing batteries used creative personnel management to meet the demand, sometimes even dispensing with the normal rule that observers should be commissioned officers.[11] B Battery, for example, was almost always short of officer observers and thus deployed two enlisted observers. SSgt. Hamilton M. Cosnahan headed the battery's team 242-Cosnahan.[12] Enlisted artillery personnel frequently had better than ordinary skills. Moreover, the U.S. Army relied on sergeants and other enlisted personnel to a greater extent than the armed forces of many other nations. Typically, however, the permanent team members were officers. Thus, on Okinawa, 2nd Lt. Ray D. Walton Jr. led one team, and 1st Lt. Samuel C. "Mac" MacCleod led another. The observation teams generally served three days on the front lines and three days back at the battery. In other artillery battalions, the procedure might be different.[13]

In contrast to the officers, the enlisted part of the team often varied from one rotation to another, and those who were not observers or heads of section sometimes skipped rotations. One of B Battery's enlisted men estimated that he spent only 10 percent of the battle on the front line even though he was a communications (signal) specialist (whose skills were in demand for observation teams). The B Battery commander and battery first sergeant drew up the team duty roster, distributing the duty throughout the battery. The high casualty rate ("turnover") on the teams during the Okinawa campaign and the desire to maintain battery morale mandated this practice. Lieutenant Thompson reported that virtually all of the casualties in the 362nd Field Artillery Battalion occurred among forward observation and liaison team members.[14] Almost every enlisted man in B Battery was at some point pressed into service on the forward observation teams. In order to keep the artillery pieces fully effective, however, no more than one man at a time from each of the four gun crews served on the forward observation teams. Only the mess sergeant, the supply sergeant, and the ammunition sergeant were exempt from observation team service.[15]

Forward observation teams typically had an observation post on the front line, although the post could be a deathtrap if the team became isolated from the infantry.[16] First Lt. Oliver Thompson noted that "the Japs were on your flanks and in your rear," and Japanese soldiers who had been "bypassed" by American troops were a threat "until death."[17] In battle, the forward observer team typically dug in near the infantry company commander (usually a captain) on or near the front line. Usually they were stationed a bit behind the infantry, perhaps with a small hill between them and the front line.

On Okinawa, the Army used two types of artillery observation teams with each infantry battalion: liaison teams and forward observer teams. The liaison team generally worked with the infantry battalion commander and was stationed slightly behind the front line. The liaison team was considered part of "the colonel's party," which consisted of the battalion leader's staff (see chart C). An artillery liaison team might also be assigned to the next-higher infantry unit—the regiment.[18]

Most liaison teams were commanded by a captain or a first lieutenant on the verge of being promoted to captain (like Charles Sheahan).

CHART C
OBSERVER TEAMS COMPARED WITH LIAISON TEAMS

Forward Observer Teams	*Artillery Liaison Teams*
support infantry company (smaller unit)	support infantry battalion (bigger unit)
led by a lieutenant	led by a captain
home unit = firing battery	home unit = headquarters battery
located on or near the front lines	located slightly behind the front lines
rotate to the rear every three days	do not rotate, remain with infantry
seek guidance from liaison officer	give guidance to forward observers

These officers frequently possessed more experience in adjusting fire than the forward observers. Thus the liaison teams were assigned to support larger infantry units. (An infantry battalion was four times larger than a company.) Beneath the team's commanding officer were three or four other men, including radio operators and a jeep driver. Sheahan recalled that four men were more or less permanently assigned to his team. Despite his rank, Sheahan took turns carrying the heavy 110 Radio with his radio operator, Wojic.[19] The liaison personnel were generally drawn from the battalion's Headquarters and Headquarters Battery—a "nonfiring" unit without howitzers. The forward observers were assigned to the "firing batteries."[20]

The liaison team monitored forward observer team fire missions and sometimes called in missions as well. Although an experienced enlisted man on the liaison team was usually capable of calling in fire, on Captain Sheahan's team only Sheahan did that.[21] If the liaison team could not see the targets, the forward observation teams called in the artillery fire, contacting the FDC at the artillery battalion directly. In such circumstances the liaison team would listen in on the fire missions to minimize the potential for error.

The liaison teams typically had two forward observation teams under them for each infantry battalion. In particular the liaison team sought to prevent friendly fire. Historian Janice McKenney described liaison teams as "a key link between the forward observer sections and fire-support resources. . . . The liaison officer's primary functions were to plan fires in support of infantry operations and coordinate target information."[22] The experienced liaison observers were expected to handle complex artillery problems. Second Lt. Donald Burrill, one of the forward observers who worked under Sheahan, recalled that Sheahan, not the forward observer, controlled the fire when it neared friendly lines. The observers could not call in 105-mm artillery fire closer than one hundred yards from friendly troops and could not call in heavy artillery closer than two hundred yards. On the rare occasions when they called in air support, it could be no closer to the troops than one thousand yards. Getting closer than that required permission from the infantry.[23] Chart D explains which types of artillery units supported which types of infantry units.

Although it might seem logical that the observation teams along the front line would draw more enemy fire than the battalion liaison teams further back, this was not always the case. For example, the Japanese typically avoided firing their heavy 320-mm mortar shells too close to their own lines. Although these "spigot mortar" rounds created massive explosions, they were also wildly inaccurate and could subject Japanese forces to friendly fire. On one occasion liaison officer Sheahan participated in an American assault up a steep escarpment—perhaps Hacksaw Ridge. The Japanese responded. First, his party came under intense fire from "all kinds of [Japanese] machine guns." Next the Japanese began to bombard his contingent with spigot mortar shells "the size of an ash can." The enemy fire brought Sheahan's party to a standstill. "We were all hiding between rocks," he recalled decades later. "People [were] getting hit near around us." One of his observer teams under Lieutenant Walton was about sixty feet closer to Japanese lines, apparently positioned in dead ground shielded from Japanese machine-gun fire and too close to Japanese lines for spigot mortar fire. "I don't know what he was doing, but he was moving forward," Sheahan later said. "I wasn't calling [in artillery] because I couldn't see anything . . . but he evidently could because he was higher up . . . near the top. . . . I wished him good luck."[24]

The distinctions between the duties of the liaison team and the duties of the forward observation team were not always maintained in combat.

CHART D
ARTILLERY SUPPORT FOR INFANTRY UNITS (WITH EXAMPLES PERTINENT TO THIS CASE STUDY)

Artillery		*Infantry*
battalion (361st Artillery)	supports	regiment (381st Infantry Regiment)
HQ Battery (Headquarters Battery) (liaison teams)	supports	battalion (2nd Battalion of the 381st)
firing battery (Battery B) (forward observer teams)	supports	rifle companies (Companies E, F, and G)

Captain Sheahan recalled significant overlap and considered his job as a liaison officer roughly equivalent to the role of the forward observer. "They were the same thing," he recalled. "Basically the same thing." Liaison officers might be pressed forward because of the shortage of observers. They might also move up to the front line because the infantry colonel elected to position himself there. "Well, I used to hang around with the battalion commander," Sheahan recalled. "He was always up front. . . . To tell the truth, mostly I was the forward observer."[25]

The breakdown in team distinctions had consequences in some units. Generally the forward observation teams were expected to absorb more casualties than the liaison teams, and that was true for the 361st Field Artillery Battalion. In a sister artillery battalion, however, the liaison teams absorbed higher casualties. Two liaison officers supporting the 1st Battalion of the 382nd Infantry Regiment were killed in action, and a third was severely wounded. Although standard procedure called for an artillery captain to be assigned to this slot, no one wanted the assignment. "The job was offered to other first lieutenants but they declined," recalled Lieutenant Thompson. So Thompson got the job.[26]

Distinctions between officers and subordinates tended to fade on the battlefield. Frontline artillerymen often disregarded the traditional customs of military courtesy, with the Army's blessing. Indeed, the Army mandated that some formalities of rank were to be ignored on the battle line. "It was a court-martial offense if you called an officer by rank," recalled Private Moynihan. They called each officer by his last name. Captain Sheahan characterized the relationship between team officers and men as "palsy walsy" (although none of the enlisted men called him by his first name).[27] Officers sought to conceal their status because they were prime targets for the Japanese. They customarily kept their binoculars underneath their uniform blouse so the Japanese could not see them. "An officer would never wear his insignia on his hat," Moynihan remembered, and "no insignia on uniforms," adding that the Japanese "would shoot for rank." The higher-ranking personnel "would be the first ones to get hit." Officers in Europe put tape over their insignia so that enemy troops would not target them in combat.[28]

Back at the battery, more traditional military customs applied. Such customs frowned on fraternization between officers and enlisted men. There were exceptions. Lt. Al DeCrans—a former sergeant—did not always comply with that rule. Although he got along well with the battery officers (whom he admired), DeCrans spent more of his time associating with the enlisted men. He frequented the kitchen and chatted with the cooks and other enlisted men, including observation team members, whom he knew far better than he knew the officers. And he continued to do that even when B Battery's first sergeant objected. DeCrans viewed the battery's senior enlisted man as a decent individual and a good soldier, but he was regular Army.[29]

As specific targets, the frontline artillery teams tried to maintain a low profile on the line of battle. They resented visits from outsiders—even high-ranking ones—to their prime viewing locations, aware that the Japanese were always watching. The case with Col. Edwin T. May illustrates the point. Colonel May commanded the 383rd Infantry Regiment, which was frequently positioned adjacent to the 381st. He often visited DeCrans' forward observation post on the Shuri Line, standing in plain sight as he examined the view with binoculars. Within two or three minutes the observation post would be receiving nasty Japanese fire. That did not affect Colonel May, however, because he left after about thirty seconds. DeCrans nevertheless recognized that May—who was killed later in the Okinawa campaign—was a good officer.[30]

Even tight-knit units—whether families or observation teams—sometimes squabble. Personality conflicts occurred on the front lines, albeit perhaps less frequently than back at the battery. Men from different backgrounds were pressed together in tense combat situations. In some cases, resentment lingered for decades after the battle. Private Moynihan recalled one such situation. The second-highest-ranking person on his liaison team was a Sergeant "Green" from Louisiana. Sergeant Green was the "head of section," but in Private Moynihan's opinion, "he didn't know nothin' from nothin'. He should have never been a sergeant. He couldn't even operate a radio."

> When he used to start to give me a bad time, . . . I would say, "Now [Green], I've got a message for the colonel and . . . I want you to take it over to the colonel." And he says, "I'm not going to do it, I'm a sergeant!" I says, "OK, you take over the radio, I'll go do it." And he says, "OK, I'll go." And he had to go because he couldn't operate the radio. . . . And that went on all the way through Leyte and all the way through Okinawa. He used to sit and smoke cigarettes and that's all he did!

Moynihan felt that almost everyone else in the liaison section "got along fine," including "the captain because he and I did all of the work."[31]

The cave-in endured by 2nd Lieutenant DeCrans illustrates the close-knit observer team relationships. DeCrans and his men were on the north slope of Conical Hill on the night of May 25–26, 1945, supporting Capt. Willard Bollinger's F Company on the front line. The U.S. infantry had already seized the summit. The cone-shaped peak stood directly in front of DeCrans' team. Even though the Japanese no longer held the peak, they were still entrenched on the reverse slope. DeCrans' team occupied a position previously held by the Japanese. "It was a cave-like hole that the Japs had dug in the north side of the hill," DeCrans recalled. The position had provided the Japanese forces with "good cover and good observation when they were defending the hill," and the observation team planned to use it for the same purpose.[32] DeCrans' team dug a neat, round opening for the cave into the side of Conical Hill. As usual, he was teamed up with T/4 Archie Hall and Sgt. Jack Gilday.[33]

The three soldiers typically split up the watches at night. Gilday and Hall changed the watch about 2 a.m. and were thus both outside the sleeping hole when a Japanese shell exploded above the cave. Although DeCrans did not hear the sound, he thinks the blast probably woke him up. He heard Gilday and Hall hollering at him to get out of the hole. He was almost out when the cave collapsed, leaving only part of his face outside. He could not breathe. His teammates uncovered his mouth and nose, but the mudslide had pinned DeCrans from his shoulders down to his legs. He could move his legs, which were still inside the hole, but his arms remained pinned by the dirt. Gilday and Hall went to work, finally freeing him after fifteen or twenty minutes of intense labor. Then and today, Gilday and Hall are heroes in DeCrans' eyes.

To the astonishment of Gilday and Hall, after DeCrans was freed he began digging to reopen the cave. DeCrans explained that his .45-caliber pistol remained inside. The two men told DeCrans to forget about it. Then DeCrans explained that pictures of his girlfriend (later wife) were in the cave. The two men pitched in to help DeCrans open the collapsed entrance. DeCrans recovered both his .45 and the photos.[34]

The story of DeCrans and his team offers important insights into the lives of forward observers. Normal hierarchical structures of rank and status broke down when men were living in the same hole in the ground. Officers and enlisted alike were reduced to their simplest forms—equals who depended on each other for survival and could die in any number of extraordinary ways. Observation team members—regardless of rank—were interdependent during combat. DeCrans' narrative establishes a point about the Japanese use of artillery on Okinawa as well. Although Japanese artillery rarely made use of forward observers, the gunners made ample use of prearranged fire.[35] Indeed, the Japanese had planned a defense-in-depth by creating a series of zones on which they could concentrate artillery fire. Like the French in World War I, they vectored in their own positions, anticipating that they might be overrun and that the Americans might eventually make use of those positions. Even outclassed and outnumbered, the Japanese artillery could still punish American infantry and the forward observers who accompanied it.

DeCrans' story also illustrates some of the major differences between liaison teams and forward observation teams, beginning with the rotation policy. The liaison officer and the liaison team remained near the front during combat and did not rotate back to their home battery, which was some distance behind the front lines. They stayed with the infantry battalion. "I would be there almost all the time," Captain Sheahan recalled.[36]

For the observer teams, rotation meant readjusting to new tasks. One team did not have to stay in position until it was relieved, but usually there was no more than a fifteen-minute gap between one team leaving the front and the replacement team arriving.[37] As soon as the forward observation team reached their home battery, the team was dissolved and the team members took on duties appropriate for their rank and the needs of the battery. Walton, for instance, recalled the contrast from the front lines, "where we didn't even take our boots off at night," to a daylight-hours-only job of directing the fire of the battery's 105-mm howitzers.[38]

Not all enlisted men were eager to return to the battery. K. P. Jones, a corporal on an observation team in Europe in 1944–45, perceived battery life as "dull and monotonous duty, such as pulling guard and digging in." When asked if the relationship between the officers and men changed when they came off the front lines and went back to the battery, Private Moynihan remarked, "Well, they made you do guard duty." Each howitzer that was not firing had to be guarded by one person at a time in rotating two-hour shifts. There had been little Japanese artillery fire against American batteries during the Leyte campaign. Each U.S. artillery battalion had been able to bunch together to form night defensive positions, and that limited the number of guards needed. The effective Japanese counter-battery fire on Okinawa meant that the guns had to be more widely spaced. Each battery had to guard its own expanded perimeter "against night infiltration attacks." More men were needed for guard duty, and weary artillerymen like Private Moynihan were awakened in the middle of the night to stand watch.[39]

One might assume that men would prefer the tedious routine of the rear area to the danger on the front lines. Artillery telephone operator Roman Klimkowicz certainly did. But Private Moynihan often felt safer when stationed with the infantry than when he was back at the battery, which was not as well dug in as the frontline infantry. Other artillerymen felt the same way. When Moynihan was with the infantry, he was securely dug in and had the infantry right there between him and the enemy. The frontline infantrymen were less jittery at night because they were accustomed to frequent night attacks. The artillerymen at the battery were only rarely exposed to such attacks and were nervous about the possibility. B Battery's proximity to the front lines may have been responsible for some of the artillerymen's concerns. Captain Rollin Harlow, the B Battery commander, wanted to be near the action and kept the battery moving forward. Staff Sergeant Knutson, who commanded one of the battery's four howitzers, recalled that sometimes the battery was so close to the front that Japanese small arms fire bounced off the camouflage poles on the battery perimeter.[40]

Back at the battery, enlisted men confronted a more entrenched military hierarchy and the potential for friction between officers and enlisted men was greater. Personality clashes loomed larger. Private Moynihan, for example, disliked Major "Smith," one of the executive officers at the 361st Artillery Battalion headquarters. Behind his back the enlisted men called him the "soap salesman" because that had been his civilian occupation before the war.

> I didn't like him—nobody liked him because he was very egotistical and he thought himself very important. And I can tell you this, what he did—one of our [unintelligible] fellows was a machine gunner and he made him go out and string hand grenades around the battery. And he made him go out and put the pins in every morning and take the pins out every night. And one of them [grenades] exploded and the fellow lost both his arms. Not dead but he spent the rest of his life in a . . . veteran's hospital.[41]

Captain Sheahan recalleded Major Smith in less than favorable terms as well. When Sheahan was first assigned to the 361st back in the United States, he and his wife had arrived late at night, and there was no one at headquarters. Sheahan, then a lieutenant, waited awhile but then left and found a place to stay for the night. "In the morning he gave me hell for not waiting longer. Yes, he was autocratic." In the major's defense, it should be said that the executive officer in battalion-level Army units is often put in the position of being the "tough guy" because he is responsible for seeing that the commanders' directives are carried out and for managing the staff officers. The commander can then be the "nice guy" father figure.[42]

Battery life, however, was by no means unbearable for members of the 361st on Okinawa. The popularity of the battalion commander, Lt. Col. Avery Masters, far outweighed the unpopularity of his executive officer. Sheahan described Masters as "very good, a small fellow, you know, not very tall, a gung ho guy." Lieutenant Walton likewise remembered Colonel Masters favorably as a fellow engineer and as a resident of Utah, the home state of Walton's wife.

The fading distinctions between officers and men on the front line seemed to have no effect on the artillery's functioning. American artillerymen performed their duties both by day and by night. Infantry company commander Bollinger, for example, considered that the artillery did "a hell of a job in front of me."[43]

The U.S. Army offensive on Okinawa relied heavily on artillery. George Feifer was only slightly exaggerating when he described the tactical doctrine of the 7th and 96th Divisions as taking "one's time, blasting and blasting again with every available shell before advancing with bodies." Marine rifleman Jim Boan likewise observed, "It was well known that the army preferred to soften up the enemy with artillery before sending infantry forward."[44]

American tactics on Okinawa focused on daylight action. The observers normally stood near the line of battle, and the selection of a good observation post was vital. Close coordination with the infantry company commander was likewise critical. When the team was with infantry captain Bollinger's F Company, for example, the forward observers situated themselves close to his command post day and night.[45] Sometimes artillery observation teams were expected to participate directly in daylight infantry attacks. Some infantry commanders even had a standard operating procedure of leading the attack on the next objective themselves, with their field artillery observer in tow. Lieutenant Walton described one such leader. "Captain Reuter [commander of G Company, 381st Infantry Regiment] . . . would get machine-gun fire placed on the area ahead of us; and then he, his radio operator, the field artillery forward observer, and his radio operator would form the point and lead the way up to capture the ridge, requesting infantry fire or field artillery fire as needed on the way."[46]

Forward observers could call in fire support not only from their home battalion's battery but also from larger-caliber division and corps-level artillery pieces. If the situation required it, additional resources, including naval and tactical air support, might be made available.[47] The 96th Division's artillery headquarters "controlled the fire of a large number of warships and . . . coordinated the bombing and rocket missions of aircraft."[48] Thus in rare circumstances, a frontline artilleryman could also coordinate bombing from naval aircraft and surface gunfire. Low-level Army forward observers did not, however, routinely direct air and naval strikes. Typically, specially trained people handled such fire missions.[49]

On Okinawa, American tactics changed when the sun set. Aggressive by day, the Americans adopted a defensive posture at night. "The Japs fought at night," Captain Bollinger said, "we didn't." Marine rifleman Boan recalled that he "and other Marines dreaded the night and prayed for dawn; the tension would not go away."[50] Bollinger remembered the shrill cries and imperfect English of Japanese soldiers in the night shouting, "Kill Americana!" After dark, the Americans generally shot at everything that moved.[51] The moving objects were all too frequently Japanese soldiers. Occasionally, however, the things that moved in the night were farm animals or civilians, including women and children.

Failure to comply with the "shoot first" philosophy sometimes had negative consequences. Captain Bollinger lost a first sergeant named Ward one night when someone forgot to frisk prisoners coming into the company area. Although the prisoners had come in under a white flag and in civilian clothes, one of them had a grenade and threw it at Ward in his foxhole. After being deprived of the highest-ranking enlisted man in their company, the men in Company G were thereafter skeptical of prisoners and white flags. Taking prisoners became difficult for Bollinger even though the rear-echelon officers wanted them for interrogation.[52]

Night defensive fire constituted a critical artillery service. The Americans fired some of their largest artillery concentrations after dark, when Japanese soldiers felt more confident leaving their underground fortifications. Lieutenant DeCrans recalled a massive gun concentration on a suspected Japanese bivouac area spotted by an air observer. The next morning, no Japanese were seen moving in the area at all.[53]

Prior planning constituted an important aspect of night defensive fire. Private Moynihan recalled how it was done: "The forward observers would lay in defensive fire and then every night my captain [the liaison officer] would go and lay in defensive fire. . . . [H]e knew where they were, and then all during the night they would call it by number. So we're just ready to down a number and they would fire on that particular spot. . . . [Captain] Sheahan . . . would go up on the line and also be like a forward observer and he would lay in defensive fire for the night."[54]

In formulating the company defense plans, Captain Bollinger tried to "build in" the artillery forward observers—that is, position them in spots where they could use artillery fire to repulse Japanese night attacks. Bollinger recalled that the observation team members would usually dig foxholes a few feet away from each other so that a single grenade could not kill all of them. Lieutenant Walton, who frequently served as an observer with Bollinger's company, noted that "in the field artillery position we fired around the clock, whenever there was a need, especially during the daytime. At night we had defensive fire, so we were usually firing the guns every half hour."[55]

Artillery provided another important service at night by sending up flares that could turn darkness into daylight. General Buckner commented in a letter to his wife: "The thunder of artillery continues all day and all night. After dark the sky over the enemy is . . . illuminated by parachute flares[,] and the constant flash of big guns and shell explosions keeps flickering like distant lightening [*sic*]. This drives the enemy into the ground and gives him very little rest."[56]

Captain Sheahan used star shells fired by Navy vessels to illuminate the battlefield in front of his battalion's line and prevent the Japanese from forming up in the dark for a night attack. Sometime after the fight for Kakazu Ridge, Sheahan spent the night in a recently captured Japanese machine-gun bunker overlooking the next hill and facing an enemy bunker there. "And when the star shells—we'd look out on another hill . . . where a machine-gun was—and I could see about four bodies of Americans . . . down at the bottom. They had been shot down and the—very eerie! Very eerie! Those star shells were really—I don't know—it's not like day but it's a weird type of light. And all the bodies were there."[57]

Few artillerymen engaged the enemy without any respite. War is 5 percent terror and 95 percent boredom, as some wit remarked. The frontline artillerymen spent much of their time doing things other than directing fire. Moynihan constantly monitored communications, unless he could talk a nearby infantryman into listening while Moynihan took a nap. "I slept with the radio on and an earphone—twenty-four hours a day. I don't ever remember missing a call."[58] The frontline artillery personnel even found time to catch up on the news. Lieutenant Walton remembered that they occasionally received *Stars and Stripes*.

Frontline artillerymen sought out or created semi-secure places to spend the night. When the infantry moved forward, the artillerymen typically moved forward as well. For the observation teams, this typically meant digging a new defensive position. The artillery liaison team was stationed with battalion headquarters a hundred yards or so behind the front lines and generally moved into positions recently vacated by the advancing infantry. In fact, one artillery liaison radio operator recalled digging only one foxhole on Okinawa. Captain Sheahan recalled that his team usually did not sleep in the same foxhole. Everybody dug their "individual graves" five or ten feet apart. Now and then they found caves or tombs that they were able to use instead of digging foxholes. During the rainy season they used pup tents.[59]

The artillerymen pressed on toward the ancient capital of Shuri. Despite the blurring of distinctions between observers and liaison officers, and notwithstanding the informal relationships between officers and subordinates, there was no reduction in the competency of American gunnery. On ridges with names like Tombstone and Hacksaw, the threat of the enemy substituted for the wrath of higher authority as a mechanism for maintaining discipline and job efficiency.

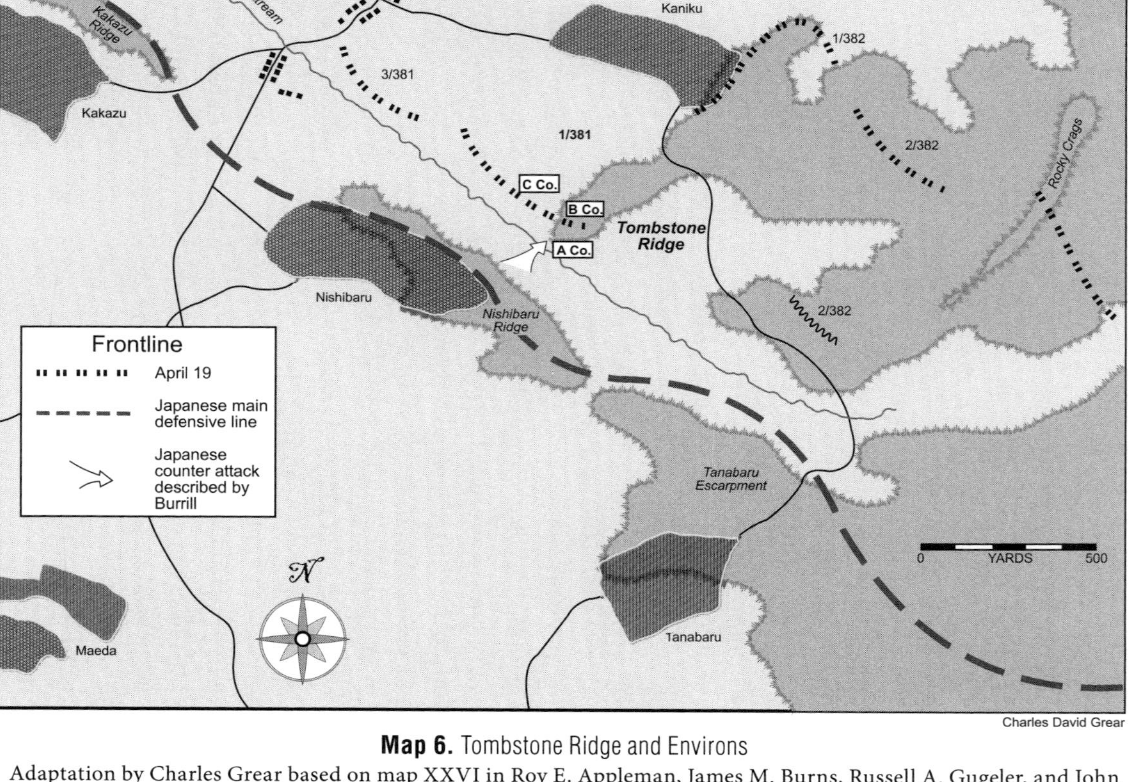

Map 6. Tombstone Ridge and Environs

Adaptation by Charles Grear based on map XXVI in Roy E. Appleman, James M. Burns, Russell A. Gugeler, and John Stevens, *Okinawa: The Last Battle* (Washington, D.C.: Center of Military History, 1948, 1991). This map is based on Lieutenant Donald Burrill's verbal account, which was at variance with the written records.

CHAPTER 6

APRIL BATTLES

In mid-April, the Americans attempted new tactics to break the stalemate on Okinawa. The 27th Division replaced the Deadeyes deadlocked with the Japanese along Kakazu's summit, and the 96th Division shifted its center of gravity eastward. Tombstone Ridge, near the center of the American line, now stood as the major geographical point in the Deadeyes' area of operation. To the naked eye the ridge was deceptively unimposing. Although it stretched about a half mile from north to south, it was only about seventy-five feet high. The Japanese, however, had strongly fortified it. As long as the Japanese held Tombstone Ridge they impeded the American advance south along Route 5 toward Shuri. Thus, despite Tombstone's lack of elevation, Appleman's team of historians it described as "the dominating terrain feature of the vicinity."[1]

Artillerymen supported the American attack, which had begun during the first week of April. An observation team led by 1st Lt. Oliver Thompson accompanied the first battalion of the 382nd Infantry Regiment. Standing at the front line before Tombstone, they witnessed Lt. Gen. Simon Bolivar Buckner Jr. using his huge binoculars to survey the "hard core of the Shuri Line." Once again the Japanese resisted the Americans' assault. Thompson's artillerymen saw a huge explosion near the village of Kaniku when a Japanese 320-mm mortar shell landed nearby. The impact threw rocks high into the air. One rock, Thompson recalled, "hit an infantryman's helmet and drove a piece of steel into his brain killing him."[2]

Out of the chaos of battle came a cry for assistance. Thompson received word that an adjacent infantry company urgently needed a forward observer. Braving Japanese machine-gun and rifle fire, Thompson and his team dashed across the battalion front to reach a beleaguered infantry platoon pinned down by Japanese mortar fire. The platoon leader pointed out the suspected position of the mortar. Thompson promptly adjusted artillery fire onto the position. As the American rounds pounded the Japanese position, the infantry platoon leader suddenly cried out, "Did you see that?" Thompson had indeed observed the impact of artillery on Japanese forces. "It looked like a Jap had been blown into the air about ten to fifteen feet," he recalled more than six decades later. The infantry leader slapped Thompson on the back. Thompson's head of section had been jumpy about coming to the front line for the first time but had performed well under fire, and Thompson suggested that the infantry officer recommend a Bronze Star for the corporal. The infantry lieutenant agreed and added a recommendation for Thompson as well. It would be Thompson's second award; he had received the same medal for valor on Leyte.[3]

Tombstone Ridge fell in mid-April 1945.[4] The Deadeyes were thus in a position to cross the rushing creek that lay just south of the ridge. Once across that stream, they could attack Nishibaru Ridge—an east–west extension of the Kakazu escarpment. The 27th Division, on the Deadeyes' right, could attempt to circumvent Kakazu by pushing down Route 5, but the ridge was directly in the 96th's path. The Japanese meanwhile mounted a counterattack to recapture Tombstone.

Three understrength infantry companies from the 381st Infantry Regiment stood in position to block the Japanese counterattack. B Company—central to the account that follows—was atop one section of the ridge. A Company had edged forward to the bottom (toe), near the spot where a road crossed the gorge on the east side of the ridge. C Company was positioned on the other (west) side of the ridge.

Second Lt. Donald Burrill typically provided artillery support for B Company of the 1st Battalion and worked in close cooperation with that unit's commander, Capt. John E. Byers. By mid-April 1945, B Company's strength was down to about fifty infantrymen.[5] Burrill's problems began even before he and his team walked up to B Company's position.[6] Typically, four other men accompanied him on a forward artillery observation team, but that day his team consisted of only three men. Burrill had just lost a very good radioman. A victim of Japanese mortar fire, the soldier had died in Burrill's arms. On his arrival at the line of battle, Burrill preplanned artillery fire to defend the company's position, concentrating his attention on the enemy positions south of the ridge. Later, Burrill left his observation team with B Company on top of the Tombstone Ridge line and walked down the slope to arrange fire support for A Company, which stood in the exposed forward position just below and northeast of the ridgeline.[7]

Frontline artillerymen sometimes found themselves in dangerous locations at nightfall, and such was the case for Burrill. Night fell before he had finished planning night fire for A Company, so he decided to remain with A Company at the foot of Tombstone Ridge rather than risk walking up the hill in the dark. The rest of his team was still with B Company on the heights. Burrill telephoned his men atop Tombstone that he would spend the night in the forward position and rejoin them in the morning.[8]

Not long after that, Japanese activity began to intensify in front of A Company. Burrill recalled that "there were Japs prowling in the dark in front. . . . [The Japanese] heard the name 'Chief' mentioned." Everyone called A Company's first sergeant "Chief" Robertson because of his Sioux blood. Some English-speaking Japanese attempted to capitalize on this situation. One yelled, "Chief, come forward and get me before they kill me." Robertson hollered back "You —— Japs aren't fooling me! You got your education in the States and then went home to help the Japs fight us. You come in here if you're so damned brave!" No one came in. Burrill directed some night artillery fire onto the area where he thought the voices had originated. He heard no more voices that night.[9]

Burrill faced yet another challenge the next morning. During the night, Japanese sappers had infiltrated the two-hundred-foot gap between B Company up on Tombstone Ridge and A Company below, and "had strung light copper wire back and forth across the path" to suggest that the trail was booby-trapped. Burrill decided to take a chance. "I got out my lineman's pliers and cut wires ahead of me," he recalled. "Every time I did this I thought a booby trap might go off but, it didn't." He made it to the top and began his observation duties. At midmorning that same April day, the Imperial Japanese Army struck the Deadeyes along the Tombstone Ridge line. From his position on the heights, Burrill saw the Japanese across the creek below him preparing to attack. The shortage of American troops made artillery assistance imperative. "We didn't have enough riflemen to hold it without help," Burrill remembered.[10]

Preplanned artillery fire proved highly effective that day. From the small plateau along the top of the ridgeline, Burrill used the night defensive fire concentrations he had calculated previously and received immediate artillery support as the Japanese units started forward. "Shoot concentration three-oh," he ordered, and relaxed a little in relief when the fire direction control center answered back: "Number one on the way."[11]

Burrill's ability to bombard the enemy without shifting fire—that is, adjusting the range—constituted a key component in American artillery success that day. If the concentrations had not been prearranged, the Japanese would have seen Burrill adjust fire and would have had sufficient warning to back off before Burrill could direct the artillery fire precisely onto the target. As it was, the artillery fire came down on the Japanese "from nowhere" as they advanced in the open. Burrill could not remember how many rounds of artillery he had ordered that day. At times he fired all four guns from his battery. For the most part, his targets were less than five hundred yards to his front. Some Japanese riflemen got as close as one hundred yards. "None of it was from very far," Burrill remembered. "They had to come uphill to get to us. And of course, that made them easy to see. When they got about so close, you didn't dare let them get any closer. You better let them have it."[12]

Forward observers on Okinawa were sometimes compelled to make quick decisions about artillery fire techniques. When the Japanese advanced in waves toward B Company's position, Burrill shifted the artillery fire "right and left out in front . . . and let them walk into it." Circumstances also dictated immediate decisions about the type of fuse for the artillery shell. Burrill normally requested "fuse quick," which meant that the shell exploded only when it hit the ground. He considered that fuse "plenty good" for the type of battle in which he was engaged.[13]

As controllers of such powerful weaponry, the observers were high-priority targets for enemy fire even when all seemed quiet. Burrill stood up during a momentary lull in the action at around 11 a.m. and a Japanese sniper's bullet ripped through his pelvis. "I hit the ground," Burrill recollected. "I thought, 'I suppose I will be a cripple for life,' but I wriggled my toes and they moved so I felt better about it." The infantrymen, concerned about the potential loss of one of their most powerful instruments of defense, rushed to help him. There was nothing they could do, though, so Burrill "crawled back to my shooting spot and went back to work. The infantry boys piled rocks around me so I wouldn't get hit again."[14] In his recommendation for the Silver Star, the infantry company commander noted that "although painfully wounded [Burrill] continued to repeat sensings and pick out new targets and get fire on them."[15]

Ground-based artillery observers on Okinawa sometimes had the good fortune to serve as part of an air-ground observation team effort. That day, Burrill was lucky enough to have assistance from the air. The weather was bright and sunny, allowing good visibility, and light Cub aircraft piloted by trained FOs buzzed overhead. Burrill recalled that "the Cub pilots were sitting up there and picking out stuff to hit that we couldn't see." Wayne Welch, an artillery liaison officer who was one of the Cub pilots, heard about Burrill's wound and sent a message: "Don't worry, you get the close ones, we'll get the far ones."[16]

The Japanese forces were more determined than ever to eliminate the American ground observer and thus stop the brimstone that was raining down on them. They took cover on the south side of the creek, which had not been as heavily shelled as the north side, and laid down mortar fire on Burrill's observation position on the ridgeline. Burrill put an end to that by calling in artillery fire that periodically raked the creek bed "pretty good."[17]

Burrill's situation near Tombstone on that April day was unusual. Infantry and artillery commanders generally tried to supply one forward observer to each rifle company, but on this occasion Burrill was obliged to direct fire in support of two infantry companies.[18] A Company, down on the low ground north of the creek, withstood a Japanese attack from around daylight to 4 p.m. because the company had a field of fire for their machine guns.[19] As for B Company, its commander, Captain Byers, asserted that Burrill's "determination to give us support . . . let us hold the hill and cut down our casualties."[20] The Cub pilot gave Burrill's fire direction credit for one thousand kills. Burrill recalled that "we got a pretty good chunk of them. They said the side of the hill was covered with them—all the way down [the] hill and clear into the creek bottom. So we had a pretty good kill."[21]

Combat situations on Okinawa sometimes compelled observers to remain in action long after they should have been evacuated, as was the case for Burrill. Indeed, B Company's commander reported that "Lt Burrill stayed in his observation post until his mission was complete and had to be pulled out from his position."[22] At length, however, the wounded lieutenant was finally relieved. Capt. Charles Sheahan, the battalion artillery liaison officer, replaced him at about 4 p.m. during a lull in the Japanese attack. Burrill remembered that Sheahan ordered him to "get the hell out [and said he] would take my place until a replacement got there." Sheahan had been with C Company on the other side of the Tombstone Ridge line, but he was well informed about what was going on. Burrill's wounds ended his active service on Okinawa. He hobbled down the hill using an M-1 rifle with a bayonet as a crutch. The medics cleaned his wound, and the next day he boarded the hospital ship *Mercy* for the long trip to Saipan. Shipboard medical personnel stripped off his bandages and discovered gangrene. He spent seventy days recuperating.[23]

The battle that raged on the appropriately named Tombstone Ridge was not the only one that tested the mettle of the observer teams as the month wore on. If General Buckner's repeated thrusts against the Kakazu Ridge line gave him the reputation as a hard-driving, direct-assault infantryman, the Navy brass was compelling him in the same direction. The American effort against Okinawa quite logically fell under the ultimate jurisdiction of Navy admirals. U.S. forces were utterly dependent on naval support. Without open sea-lanes, no ammunition (or any other supplies) would reach the gunners on the island. The Navy, however, paid a high price to keep the supply lines open. As infantrymen slugged it out with the Japanese on land, a massive air-sea battle raged over the waters surrounding Okinawa. Suicide aircraft and standard Japanese warplanes attacked the American fleet relentlessly. The U.S. Navy losses in both men and ships placed enormous pressure on General Buckner to push his troops forward. During a personal visit to Okinawa on April 23, Adm. Chester Nimitz, Buckner's boss, was livid. "I'm losing a ship and a half a day," Nimitz told Buckner. "So if this line isn't moving within five days, we'll get someone here to move it so we can all get out from under these stupid air attacks."[24]

Buckner acted. The entire American front line made a coordinated attack on April 24. Despite their reputation for squandering lives in banzai charges, the Japanese proved as skillful in retreat as they had in their

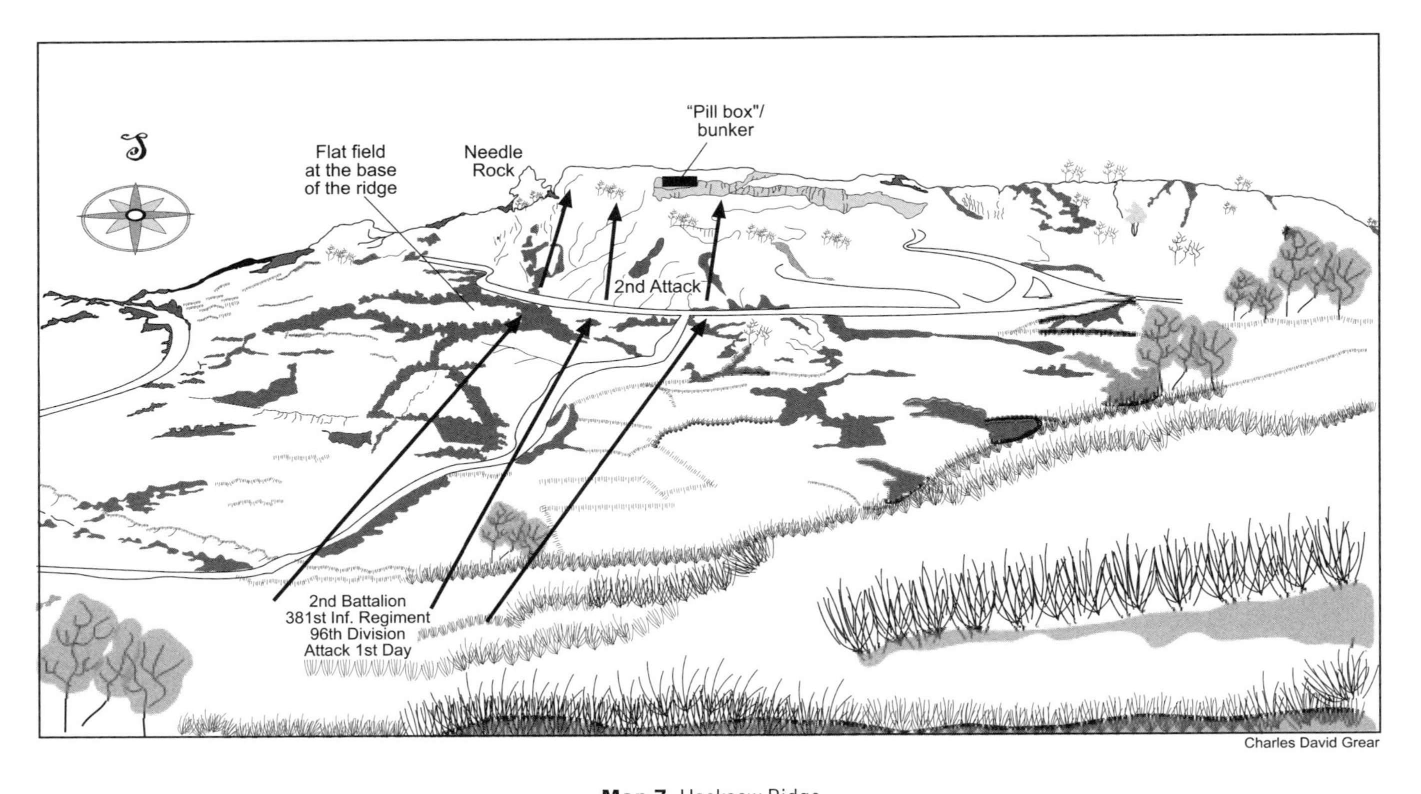

Map 7. Hacksaw Ridge

Adaptation by Charles Grear based on drawing on pp. 244–45 (Urasoe-Mura Escarpment) in Roy E. Appleman, James M. Burns, Russell A. Gugeler, and John Stevens, *Okinawa: The Last Battle* (Washington, D.C.: Center of Military History, 1948, 1991).

defense of Kakazu Ridge. Faced with simultaneous assaults from numerous units, the Japanese on Kakazu simply disappeared, leaving behind a small force to cover their retreat. Lt. Gen. Mitsuru Ushijima's forces now remained largely intact and were arrayed against the Americans about one mile to the south.[25]

The American advance on April 24 brought the Deadeyes of the 96th Division to the base of the formidable Maeda escarpment.[26] Once again the Japanese defense line consisted of high ground with a well-fortified reverse slope. The east–west-running ridge blocked forces moving south. Just past the east end of the summit of the ridge stood Needle Rock, a tall, toothlike prominence that gave the ridge its name: Sawtooth Ridge or Hacksaw Ridge.[27]

The Americans' offensive against Hacksaw's approaches on April 24 once again proved the importance of artillery on Okinawa. American artillery blasted the vegetation from the escarpment and left only bare rock.[28] After the barrage, Buckner's infantry stormed across a flat area toward Hacksaw's base, right into the formidable Japanese defenses. The Deadeyes, accompanied by their artillerymen, led the charge. Private Moynihan's artillery liaison team was in the center of the formation. On reaching a field at the base of the ridge, they positioned themselves near an abandoned Japanese pillbox within supporting distance of the infantry moving up the ridge.[29]

In this advance the artillerymen introduced a new, highly secret (until then) plastic fuse variously referred to as a proximity fuse, VT (variable time) fuse, plastic fuse, or radio-controlled fuse. Okinawa appears to have been the first Pacific land battle where proximity fuses were used to any substantial degree. The new fuses were extremely sensitive, intended to explode when they were still in flight so as to cause shell fragments to fly into enemy foxholes.[30] Their touchiness caused problems for the American artillerymen as well as the enemy. Moynihan had vivid recollections of the new fuse. "It got very sensitive," he recalled. "It could even hit rain and explode in the air over us because it was designed to be set so that it could explode before it hit the ground so that we could get the people in foxholes. And we had a lot of those! There was a lot of that."[31]

Moynihan had witnessed earlier problems with premature overhead explosions in the Philippines. In the field north of Hacksaw Ridge the friendly fire nightmare recurred when the rain caused the fuses to explode over the American troops. "I would be at one end of the line and that would happen maybe on the other end of the line, which would be far enough for me that I was protected in my foxhole. I kept my head down, I'll tell you that, because I could. . . . I always had my foxhole low as I could so I could stretch out."[32]

Other observers witnessed similar mishaps. Lieutenant Thompson recalled that artillerymen on Okinawa nicknamed the new weapon "hell fire" because of its notorious tendency to "burst way up in the air and not near the ground as in time fire." Thompson theorized that the ships' radar from the massive American fleet off Okinawa interfered with the proximity fuses and caused early detonations.[33] Lieutenant Walton agreed with Moynihan as to the cause of the premature explosions. "Those were radio-sensing fuses for 105-mm howitzer shells. They were supposed to explode, I believe, one hundred feet above the ground. . . . [W]hen they were shot, then they were armed by some centrifugal motion. The big problem, though, is that rain played havoc with them and somehow the radio waves could get grounded on the rain, I guess." Moynihan thought that this type of friendly fire problem was largely resolved after about a month (roughly two months after the campaign had begun).[34]

On April 26, the day Buckner's next offensive move began, the commanding general inspected the front lines to observe the attack on the Shuri Line (of which Hacksaw Ridge was a prominent part). He described the event to his wife.

> Yesterday I had the rare experience of finding an observation point that permitted me to observe the entire battle front. Action was lively along the line since we were putting on a heavy attack and the whole thing was visible. It was really a superb spectacle—plane strikes, artillery concentrations, smoke screens, flame throwers, tanks and the steady determined advance of the infantry closing with the enemy. Along with this were the crash of bombs, the screech of projectiles, the whistle of shell fragments, the sputter of machine guns and the crack of rifles, I shall never forget it. It was really stirring.[35]

As part of that April 26 offensive, the 381st's 2nd Battalion was ordered to move up the ridge and seize Hacksaw's summit. Sergeant William Filter's infantry squad from G Company proceeded up a steep path, meeting increasingly fierce resistance as they neared the summit. Enemy fire killed the soldier right in front of Filter. "It was hell for about an hour," Filter recalled.[36] E and G Companies needed only forty minutes to get up the ridge. Once there, however, they could not join up because a pocket of Japanese soldiers held out between them.[37]

Intense fighting raged up and down the slopes of Hacksaw Ridge.[38] Capt. Willard Bollinger considered that fight the toughest of the Okinawa campaign. His F Company was positioned on the very top of the ridge around Needle Rock. The Japanese held the reverse (southern) slope, and the Americans held the forward (northern) slope except for a reinforced concrete bunker built into the cliff.[39] The summit changed hands repeatedly. A widely circulated 1947 magazine article describes F Company's fight as a Wild West battle "complete with shots from the hip and close-in knife play."[40]

Even artillery liaison teams—artillery's second echelon—could end up in exposed positions close to the line of battle. Like Burrill in the previous battle, Moynihan's team found themselves in a bad situation. The commander of the 2nd Battalion reached Needle Rock on April 27, accompanied by his artillery liaison team.[41] Pfc. Charles Moynihan was not impressed by what he found there. "We finally moved up just to the base of the Needle Rock. . . . My foxhole was lined up almost underneath it. Down probably, oh between fifty and a hundred feet. And that's where the colonel was but he was in an awful spot. He didn't have any pillbox or tomb. . . . I didn't think too much of this.[42]

Confusion reigned near the summit. Japanese defenses honeycombed the escarpment. Evidence of that was visible even up where Moynihan was. "The infantry would throw those white phosphorus hand grenades into those caves and they would come up all around us in the air vents," he remembered. Residue from blasts inside the hollowed-out ridge also flew out of air vents, some of which were located behind the 2nd Battalion headquarters. "It would be all around us, even back of you," Moynihan recalled.[43]

One particular Japanese defensive position demonstrated some of the limitations of American field artillery fire on Okinawa. Not far from Needle Rock the Japanese had built a concrete bunker into the face of a cliff atop Hacksaw, with an opening that gave the defenders a bird's-eye view of the attackers below. This was the pillbox that prevented E and G Companies from linking together on Hacksaw's summit. As long as the Japanese held the pillbox, the summit of Hacksaw was insecure. At night, Japanese reinforcements could sortie out of this bastion. Gunnery proved unsuccessful in reducing the pillbox. Bollinger said that the shells just "bounced off."[44]

Artillery having failed, the infantry moved in. On April 28 the Deadeyes attacked the Hacksaw pillbox with explosives and a flamethrower.[45] The defenders held fast. They were reinforcing the bunker from underground tunnels. Flamethrowers failed to dislodge them. A satchel charge dropped into the pillbox caused an immense secondary explosion as if it had sent up an ammunition dump. Bollinger felt the ridge shake as if the whole mountain were going to explode.[46] Even that didn't do the trick. The weary remnants of the 2nd Battalion never reduced the Hacksaw Ridge pillbox in their midst. Even as the 96th Division departed for a well-deserved rest, the 77th Division battled on against the same pillbox for several days.

One month of heavy fighting had taken a tremendous toll on the American infantry. By April 29th the 381st Infantry Regiment had lost 56 percent of its authorized allotment of enlisted men and 32 percent of its officer strength.[47] The battle had especially depleted the rifle company's manpower, leaving only weak resistance standing between the artillery liaison teams and the enemy. Private Moynihan remembered his concern after learning the dire news on Needle Rock. "When they put the report in about how we lost the men, I know we were down to eighty-eight infantry, that was all that was out there at one time—all that was in front of us from the battalion was eighty-eight men."[48]

Relief arrived on the morning of April 29 when elements of the 307th Infantry Regiment of the 77th Infantry Division replaced the Deadeyes' 2nd Battalion.[49] When the company commander from the 77th saw only a few 96th Division soldiers leave the G Company positions on Hacksaw Ridge, he asked Capt. Louis Reuter Jr., "Where are all your men?"[50] Reuter's reply went unrecorded. G Company had been relieved just in time. One infantry sergeant considered that a company-strength banzai charge by the Japanese would have annihilated G Company's "paper thin" lines.[51]

Although the bulk of the 96th Division departed for the rest camp, Captain Sheahan had to remain. His artillery liaison team stayed to support the 77th Division. "You see," he later explained, "artillery is never in reserve. So when [the infantry was] relieved, I would stay there after the 77th relieved us. So I adjusted artillery for the 77th . . . because I had all the concentrations.[52]

The weary Deadeyes moved north into a rest camp. For approximately ten days the tired men recuperated as the division took on replacements and trained new men. One of the forward observer "replacements" was 2nd Lt. Al DeCrans, back from sick leave. DeCrans had actually joined the Deadeyes in the Philippines, where he contracted a tropical disease. Now recovered, he returned to the 96th Division in late April 1945 and joined B Battery. His FO team sometimes rotated with Lieutenant Walton's team. The 381st Infantry Regiment had been "beaten up really bad" by the time he got there, DeCrans recalled. Three thousand new replacements for the infantry units arrived from the States about the same time DeCrans arrived from the Philippines. Unlike DeCrans, many of the reinforcements arriving on Okinawa were quite green. One young soldier assigned to Lieutenant Walton's observation team had never even seen an artillery shell explode before he witnessed combat on Okinawa.[53]

Infantry command changes affected future tactics for artillery and infantry alike. The 381st's 2nd Battalion—the infantry unit that DeCrans and Walton supported—had a new commander. Maj. Leon Addy replaced Col. Russell Graybill as battalion commander on May 8, 1945.[54] Private Moynihan recalled that the new major "was very business-like and he was very focused and he really pushed the troops. He was more aggressive. He was more demanding" than Graybill.[55] One noncommissioned officer in the infantry battalion, Sergeant Filter, also described Major Addy as much more aggressive than Colonel Graybill and as a soldier who wanted to be up front. Before the battle of Okinawa ended, Major Addy's aggressiveness would cost him his life.[56]

CHAPTER 7

REDUCING THE SHURI LINE

In May 1945, at the midpoint of the battle of Okinawa, the Japanese conducted their largest counterattack of the Okinawa campaign. It began on May 4 and continued into the next day. The 96th Division had been rotated into rest camp, so for two days the brunt of the attack fell on the other American line divisions. The Americans blunted the attack, much to the dismay of Lt. Gen. Isamu Cho. Col. Hiromichi Yahara, a defensive specialist who had opposed the counterthrust, gained more influence with the Japanese commander.

The Japanese still held fast at the Shuri Line. The line centered on Shuri Castle, a fortress atop a hill in the ancient capital of Okinawa with a spectacular panoramic view of the island. The castle's vast underground tunnel system served as the headquarters of Lt. Gen. Mitsuru Ushijima, the commander of the Okinawa defense forces. Two tunnel systems lay underneath Shuri Castle: one for the 32nd Army command and one for the artillery command. Even though the aboveground structure was almost completely destroyed by naval gunfire, bombing, and artillery fire, not a single member of the Imperial Japanese Army command was killed while occupying the tunnels.[1]

Two hill masses anchored the ends of the Shuri Line. Much has been written about the western anchor—Sugar Loaf Hill—which the 6th Marine Division took after a vicious fight.[2] The American hero of that battle was Marine major Henry A. Courtney Jr., who was awarded the Medal of Honor posthumously for his valor. Conical Hill, on the east, has attracted far less attention. In many ways, however, Conical Hill was as important as Sugar Loaf and perhaps more so. Lt. Gen. Simon Bolivar Buckner Jr., the commander of U.S. forces on Okinawa, certainly thought so. He called Conical "the key to the Shuri line."[3] Conical Hill was the dominant terrain feature on the eastern side of the island, and

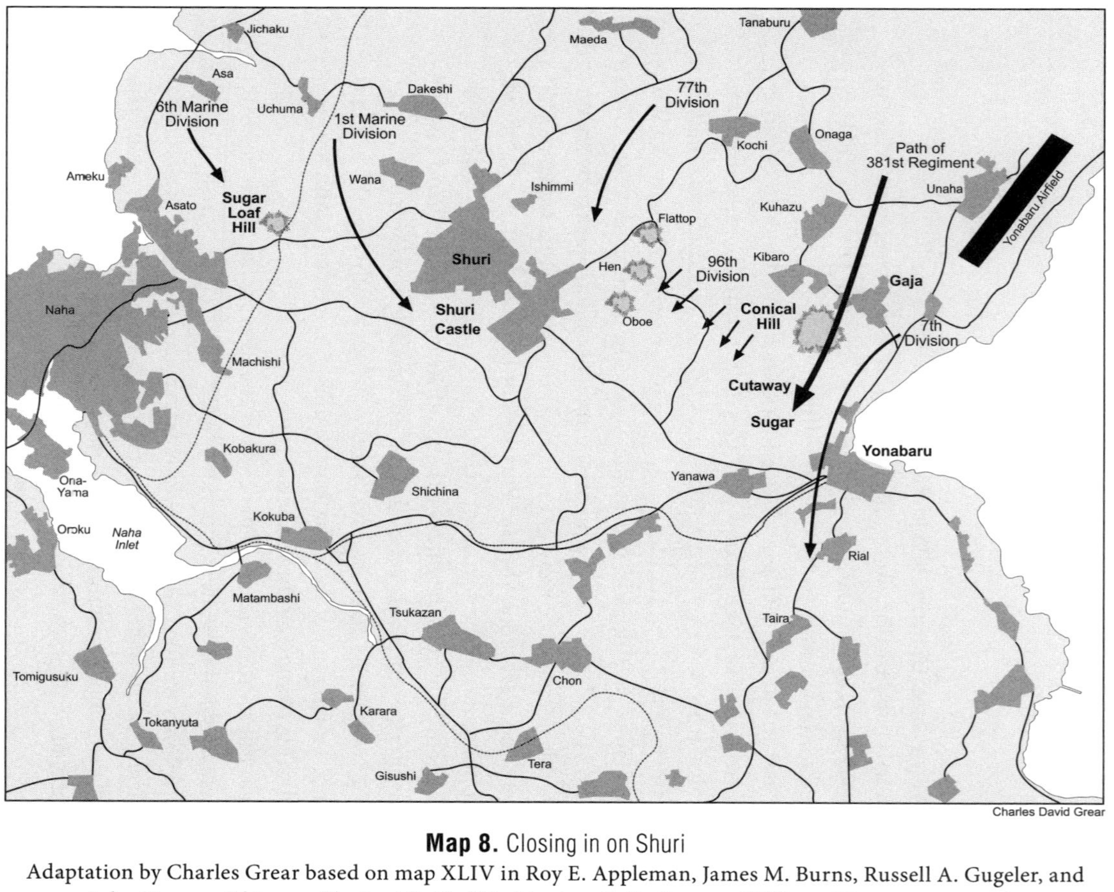

Map 8. Closing in on Shuri

Adaptation by Charles Grear based on map XLIV in Roy E. Appleman, James M. Burns, Russell A. Gugeler, and John Stevens, *Okinawa: The Last Battle* (Washington, D.C.: Center of Military History, 1948, 1991).

the Japanese defenders there had a clear view of the eastern Shuri Line and the approaching Americans. Those defenders had the potential to block the Americans' drive along the coast to the port city of Yonabaru.

On May 9, the Deadeyes headed toward Conical Hill after roughly ten days in rest camp. On that same day the 381st's two sister regiments from the 96th Division moved forward to relieve the 7th Division. That same wet, muddy May 9, the U.S. forces on Okinawa learned that Germany had surrendered in Europe. The 381st Infantry Regiment moved south along the coast in the direction of Yonabaru. On May 10 the 381st set up a new command post, taking over a sector previously held by the 7th Division. The 2nd Battalion headquarters took up a position east of Conical Peak and north of Yonabaru.[4]

Pvt. Ken Staley, a rifleman in one of the 381st's sister regiments, observed a grisly scene as his unit moved up toward the line of battle. "Along the way we passed other Army trucks lurching to the rear and piled high with dead American soldiers. What an introduction to War! We new guys stood in shock. . . . When we got off the trucks and marched to the front lines, we got the final shock—a scene straight from hell—bloated dead bodies of both American and Japanese soldiers lying all over the area where a battle had raged a few days earlier."[5] A veteran reminiscing long after the war speculated that the military authorities made sure that the new replacements saw those bodies, presumably to impress them with the seriousness of the task they were facing.

Artillery participated in this new fight, of course. Capt. Charles Sheahan remembered concentrating artillery on Conical Hill, and Pfc. Charles Moynihan, as usual, accompanied the infantry battalion commander's party as part of the artillery liaison team. At this point a JASCO (Joint Assault Signal Company) team that specialized in adjusting heavy naval gunfire joined them. Moynihan's unit did not participate in the physical seizure of Conical Peak, and he was unsure whether it rendered fire support against Japanese positions there. A saddle between Conical and the adjacent hill provided an adequate observation point, so Moynihan never climbed the peak. "We stayed there [on the saddle] and everything was done by radio."[6]

Another regiment of the 96th Division eventually took Conical Hill on May 13–15, but it was a vicious fight. Despite intense Japanese machine-gun and mortar fire, two platoons worked toward the summit. Col. Edwin May, their regimental commander, described the event as "the greatest display of courage of any group of men I have ever seen." The Japanese nevertheless continued to defend other parts of the Shuri Line and managed to block an American breakthrough for another two weeks after Conical Hill was taken. While they no longer held the peak, they continued to offer resistance nearby.[7]

The Japanese found ways to punish the Americans delivering combat support near Conical. Captain Sheahan typically dug his foxhole near the battalion command post at night. Air and naval liaison officers in the Yonabaru area positioned themselves close to the battalion command post as well. The Japanese on a peninsula across the bay used a 47-mm antitank gun to harass them. Sheahan remembered that the air corps liaison officer at Yonabaru slept in a hammock until the night a 47-mm shell "landed right under his hammock and killed him. That was a hard target cause you couldn't see where it was coming from." Sheahan called for the assistance of corps-level artillery—codenamed Carlotta—to suppress the antitank gun. Corps artillery possessed long-range 155-mm guns ("Long Toms"), which had a greater range and flatter trajectory than the howitzers. "I had perfect communication with them for days. I fired many rounds, in particular . . . across what was later to be called Buckner Bay. A peninsula ran directly east of Yonabaru. It was from there we were receiving great amounts of Jap 47-mm fire. A spent piece hit me and knocked me down. . . . I fired at flashes and possible positions. I will never forget 'Carlotta.'"[8]

The 96th Division units returning to the front lines were located near the town of Gaja, close to the coastline north of the base of Conical Hill. The 2nd Battalion began to clear the town of Japanese defenders with liberal doses of artillery fire. The infantry regiment records reported excellent results when 81-mm mortars and artillery were applied. The commanding officer of G Company said it was the best artillery barrage he had ever seen.[9]

Such results came at a cost, of course. The Japanese forces on Conical Hill killed one of the 361st's forward observers on May 13. Pvt. Curt Sprecher, a G Company infantry replacement, was nearby when it happened. G Company's forward line was positioned on the slope of Conical

Hill, in the shadow of the cone-shaped peak that gave the hill its name. The Japanese held the high ground. Sprecher did not know the observer's name, but he remembered that he was an officer. The observer came by and inquired, "How are you doing, soldier?" and then moved forward of the front lines. Sprecher turned away for just a moment, and the next thing he knew, the forward observer had been hit.[10]

The Japanese were not finished for the day. That same afternoon, perhaps within the hour, G Company lost the services of its popular company commander. Capt. Louis Reuter Jr. was shot in the head. Although he survived, the daring captain remained disabled for the rest of his life. Private Sprecher again was nearby, and he assumed that the same Japanese marksman—a sniper located about fifty yards to Sprecher's front—had hit both officers. Sgt. William Filter spoke with Captain Reuter later and learned that Reuter had gone up to take a look at a forward observation post atop a ridge in the Yonabaru area. Reuter reported that he saw the Japanese sniper and wanted to take a shot with his carbine, but the Japanese soldier hit him first, the bullet cutting a groove through his forehead that penetrated his brain.[11]

The loss of such a popular officer both affected morale and left the Americans shorthanded. G Company performed its assigned tasks after Captain Reuter was evacuated, but much of the company's spirit had disappeared. Willard Bollinger remembered his own discouragement. The unit, however, had to adjust. First Lt. Frederick Dilg replaced Reuter as G Company commander. The Japanese continued to take a toll on the officer corps. At one point, G Company would be so short of officers that Dilg was the only one in the company.[12]

The artillerymen obtained mixed results in the Conical Hill area. On May 16 the 2nd Battalion spotted an enemy observation post. Artillery fire was directed against it, and the observers believed that it had been "reduced."[13] But three men of G Company were wounded that day by friendly fire. Indeed, roughly twenty-five rounds of American field artillery landed in the 2nd Battalion area.[14] On May 20 the artillery results were much better. The 361st and other units placed "precision fire . . . on six enemy caves, 2 OPs and Art[iller]y and m[or]t[a]r positions. Observers reported destructive results on targets. Area targets included camouflaged enemy t[an]ks and a motor park. Hits were reported on these targets as well as enemy personnel. . . . Total of 52 missions and 4125 r[oun]ds were fired."[15]

Under the persistent pounding, the Japanese slowly gave way. They knew the significance of the loss of Conical Hill. Colonel Yahara affirmed it in his interrogation after the battle.

> On about 20 May it became apparent to the 32nd Army Staff that the line north of SHURI would be soon untenable. The pressure exerted upon the line from both Sugar Loaf and Conical Hill forced a decision as to whether or not to stage the last ditch stand at SHURI. The capture of Sugar Loaf Hill alone could have been solved by the withdrawal of the left flank to positions S[outh] of NAHA and, in Col[onel] YAHARA's opinion would not have seriously endangered the defense of SHURI. However, the loss of remaining positions on Conical Hill in conjunction with the pressure in the west rendered the defense of SHURI extremely difficult.[16]

For the Japanese, the situation was becoming increasingly tenuous.

By May 21 the Americans were threatening to outflank the Japanese on both ends of the Shuri Line, at Sugar Loaf and Conical Hills. Historian Thomas Huber noted that by May 21, "elements of the U.S. 96th Infantry Division . . . were close to turning the Japanese flank at Yonabaru."[17] The 96th Division was trying to hold the Japanese away from the coastal road on the eastern side of the island, a task the Deadeyes accomplished by engaging the Japanese on a series of hills that ran alongside the coastal road that went south through Yonabaru. In the meantime, the 7th Division sought to barrel down that road to secure the port of Yonabaru and then push south. The plan worked, much to the surprise of the Japanese. Colonel Yahara later admitted that "the occupation of YONABARU on 22 May came as a surprise to the Japanese who did not expect such a move during the inclement weather prevailing at that time, assuming that [U.S.] infantry would be unwilling to attack without tanks which were thought to be immobilized by the mud. On 23 May elements of the 24th Div[ision] were dispatched to retake the town. The attack continued with no success on the 24th and 25th May."[18] The Americans' advance slowed after the seizure of Yonabaru. Captain Sheahan remembered a failed U.S. night attack in the area around that port city. The unit advanced a short distance but was stopped by machine-gun fire and mortars.[19]

In the meantime, American artillery was wreaking havoc on the Japanese. A Japanese prisoner captured about May 31 revealed that the shelling had destroyed the telephone communication lines and said that "much of the communication between units was maintained by runner."[20] His interrogation report detailed the devastating impact of the big guns: "Many of the men in rear areas were wounded by American shelling, of which artillery had the greatest effect on men and material. . . . The troops are all astounded at the amount of shells used by the Americans and refuse to move about during the day, since even the slightest indication of a movement in the Japanese area would bring down a rain of artillery shellings."[21]

In late May, the Japanese abandoned the Shuri Line and staged an orderly withdrawal toward the south. The Americans now faced two problems. First, heavy rainfall and massive flooding in late May and early June 1945 bogged down their advance. Second, they needed to maneuver their forces to pursue the retreating Japanese while minimizing the threat of pre-prepared Japanese ambushes.

Private Moynihan recalled the confusion during the turn south in late May.

> We stayed at that one position [near Conical Peak and Yonabaru]. And then when they took Conical Peak and . . . Shuri Castle, we were ordered to go around and follow the road going back toward Shuri. And our troops were going along that road on the backside and I don't know how far we marched because we were supposed to relieve some other. And then they decided—they stopped and we turned around and came back. And then we went out into those rice paddies going south.[22]

Thus, rather than continuing to advance along a road running west from Yonabaru toward Naha through Shuri, Private Moynihan's unit reversed course and turned south, toward the defensive positions the Japanese had prepared in the rural southern tip of Okinawa.

Around that time, Private Moynihan remembered the heavy rains starting and the roads getting bad. "And we went out across those rice paddies. . . . I can't remember how far we got because you see at that time the rains hit. And we didn't move too much because we couldn't get supplies. And . . . the Cub [aircraft] dropped my radio batteries. . . . We only moved two or three times while that storm was going on."[23]

The rain set in while Captain Sheahan's unit was still north of Yonabaru. "We laid in the mud for a week or two. Then we went on again. . . . The roads started to turn to mud. In fact they had to evacuate the wounded through Yonabaru by boat to get them back to the rear area. You couldn't drive. . . . The Navy . . . had landing ships pick people up. . . . Nobody went anywhere. . . . We were stuck in the mud for a while."[24]

The severe weather lasted about two weeks. Captain Sheahan's liaison team had been issued a trailer (which was usually pulled behind his jeep), but the jeep couldn't pull it because of the poor road conditions. Sheahan's team had to park the trailer on the roadside as they moved up to the front line. Someone stole the trailer and his camera too. Decades after the war, Sheahan still resented his loss.

As May turned into June, the transportation network remained dismal. The 381st's report for June 2 notes that "Route 13, the only road in the area, was in extremely poor condition, containing many washouts and deep holes." On June 4 the engineers for the 381st reported filling craters and checking for mines. On June 5 the engineers attached to the 381st "continued to work on Route 13 from YONABARU to 8166-H5 (where bridge is out). Pulled vehicles through bad spots and filled holes in road. Impossible to get material to YONABARU via road or water to construct bridge."[25]

The misery was endemic. Lieutenant Walton vividly recalled the horror of a night truck movement in the mud while enduring a bout of diarrhea.[26] General Buckner described the miserable conditions to his wife but found humor in one report from Maj. Gen. "Smiling Jim" Bradley, the commander of the 96th Division.

> We continue to be deluged with rain, about seven inches having come down during the past four days. This came at an unfortunate time, since I had caught the Japs napping and shoved a

> division past their right flank. . . . While our progress is still slow the Jap casualties continue to be high. Bradley, who commands the 96th Div., was on a particularly slippery range of clay hills when the heaviest of the rain struck him. In his daily report of movement he reported: "Considerable movement by my front lines. Those on forward slopes slid forward and those on reverse slopes slid back."[27]

Lieutenant Walton's observation team rotated to the front line in these awful conditions.[28]

General Buckner's response to General Ushijima's retreat proved controversial. Military historians have criticized Buckner for his delay in recognizing the Japanese withdrawal from the Shuri Line and his slow pursuit of the Japanese retreat to the south. Gerald Astor, for example, wrote that "not until May 31, did General Buckner and his people realize Ushijima's plan, and even then the U.S. generals badly underestimated the scope and the effectiveness of their quarry's escape."[29] Buckner's reactions, however, can be better understood if the weather conditions in which his frontline soldiers were operating are considered. When the rain finally stopped, it still took a few days for the transportation network to dry out. Only then could the Americans pick up the tempo of their advance.

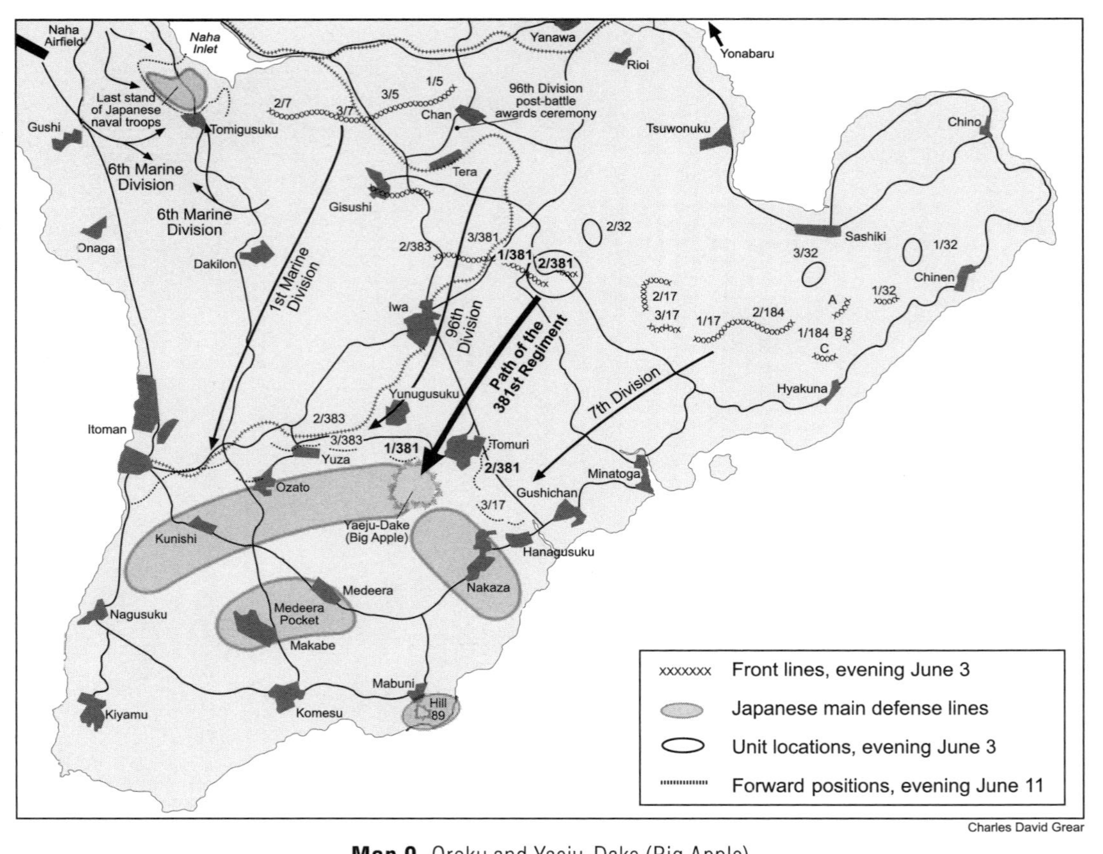

Map 9. Oroku and Yaeju-Dake (Big Apple)

Adaptation by Charles Grear based on map XLVII in Roy E. Appleman, James M. Burns, Russell A. Gugeler, and John Stevens, *Okinawa: The Last Battle* (Washington, D.C.: Center of Military History, 1948, 1991).

Okinawa, April 20, 1945: A forward observation team calling in artillery adjustments against a Japanese target. White phosphorus marker rounds explode in front of the American lines. Note that there are five men in the small observation post; two are wearing helmets, and three are bareheaded. *U.S. Army Signal Corps photograph, U.S. National Archives, 111-SC-3711 01*

Yonabaru, Okinawa, June 27, 1945: A dump for spent artillery casings. The size of the dump demonstrates the enormity of the artillery battle on Okinawa. *U.S. Army Military History Institute, 0 209936-S*

Okinawa, June 7, 1945: Lt. Gen. Simon Boliver Buckner Jr. (left), commander of the Tenth Army on Okinawa, and Gen. Joseph W. Stilwell during Stilwell's inspection trip to Okinawa. Stilwell took command of the Tenth Army after Buckner's death. *U.S. Army Signal Corps photograph, SC 207213S, U.S. Military Institute of History*

Mindoro, Philippine Islands, October 1945: Off-duty officers from the 361st Field Artillery Battalion. Capt. Charles Sheahan is at far right, and Lt. Ray Walton Jr. is standing at far left with cap. *Author's collection (gift of Ray Walton)*

Okinawa, 1945: Second Lt. Al DeCrans shaving, perhaps near the division's rear area at Ginowan. *Author's collection (gift of Ray Walton)*

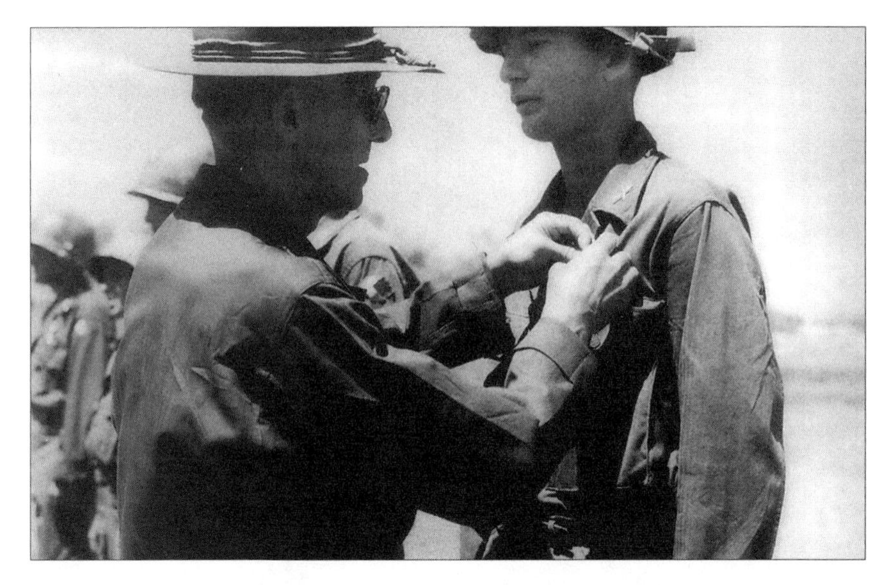

Okinawa, July 3, 1945: First Lt. Donald M. Burrill receiving the Silver Star for valor from Gen. Joseph W. Stilwell, the new commanding general of the Tenth Army. *Author's collection (gift of Donald M. Burrill)*

Okinawa, April 6, 1945: Tenth Army troops calling for artillery support. The original caption reads, "As troops of the Tenth Army advanced on Okinawa they called for continuous support by light and heavy artillery. Much of this forward support came from tanks and armored vehicles as depicted here. The gun on this half-track, having already hit the Japanese gun emplacement atop the hill in the distance, is now pouring more rounds in it to finish it off for infantrymen." *U.S. Army photograph, SC 371139, U.S. Military Institute of History, WWIISC–Geography–Japan–Okinawa–7*

Okinawa, 1993: Kakazu Gorge. Units of the 96th Division crossed this creek under mortar fire during the April 1945 assault on Kakazu Ridge. Much of the gorge is now a park. The natural creek bed of 1945 has been replaced with concrete.
Rodney Earl Walton

Southern Okinawa, May 11, 1945: Nighttime harassing fire focusing on likely avenues of approach and locations where Japanese troops might emerge from underground positions.
U.S. National Archives, NWDNS-127-N-120807

Okinawa, ca. April–June 1945: Cpl. George W. Arnold, a member of Lieutenant Walton's forward observation section, using a microphone. Note the field artillery radio on a pack board next to Arnold's right knee. *Author's collection (gift of Ray Walton)*

Tomuri, Okinawa, June 18, 1945: 7th Division soldiers observe enemy action. Forward observers and commanders could find themselves in a precarious position at an observation point. The lion-dog statue still bears the scars of the bullets. *U.S. Army photograph, U.S. Military Institute of History, CPA-45-30128, WWII S.C. Coll.–Geography–Japan–Okinawa–10*

Okinawa, ca. April–June 1945: Lieutenant Walton in military gear worn by forward observers on Okinawa. The spools of cable immediately behind Walton are for telephone lines. *Author's collection (gift of Ray Walton)*

Okinawa, ca. April–June 1945: Lieutenant Walton's observation team preparing to move to the front line. *Author's collection (gift of Ray Walton)*

Near Yonabaru, Okinawa, ca. May 1945: Lt. Samuel "Mac" MacLeod (right) briefing Lieutenant Walton as Walton's team rotates up to the line of battle. MacLeod's team was returning from the front line near Conical Hill. The bay of the port of Yonabaru is slightly visible at left, about one-third of the way from the top. *Author's collection (gift of Ray Walton)*

North of Yonabaru, Okinawa, ca. May 1945: Cpl. George Arnold observing on the front line. The stenciling on the back of Arnold's fatigue jacket identifies him as a member of B Battery, 361st Field Artillery Battalion, 96th Infantry Division. *Author's collection (gift of Ray Walton)*

Okinawa, ca. April–June 1945: An artilleryman named Fuller in a dugout under an embankment. Lieutenant Walton's team slept five abreast at this spot during at least one rotation on the front lines. *Author's collection (gift of Ray Walton)*

Near Shuri, Okinawa, June 10, 1945: Sgt. John T. Anderson (363rd Field Artillery Battalion, 96th Division) receiving a haircut from Pfc. Troy F. Dixon at a gun position amid the rubble of war. Anderson is seated in a Japanese barber chair. *U.S. National Archives, NWDNS-111-SC-208582*

Okinawa, ca. April–June 1945: Lieutenant Walton at a B Battery gun position. An Army cot is dug in below ground under the pup tent. *Author's collection (gift of Ray Walton)*

Okinawa, 1945: View seen by American troops advancing south against Hacksaw (or Sawtooth) Ridge. Needle Rock protrudes above the left center ridgeline. The photographer had his back toward Kakazu Ridge. *Author's collection (gift of Ray Walton)*

Okinawa, 1993: View of Needle Rock from the east side. Fighting swirled around this point as Capt. Willard G. Bollinger's infantry company struggled for control of Hacksaw's summit. The vegetation has flourished in the years since. *Rodney Earl Walton*

Okinawa, 1993: Shuri Castle (reconstructed after the war), the center of the Japanese main line of resistance. Although the castle was reduced to rubble during the invasion, tunnels underneath the castle protected the Japanese 32nd Army headquarters, including the artillery command. *Rodney Earl Walton*

Okinawa, May 6, 1945: Conical Hill before it was taken by the Americans. The hill commanded the surrounding countryside and anchored the far-right wing of the Shuri Line. The small body of water at the far left, center, is Yonabaru harbor. *U.S. Military History Institute, CPA-45–4472, WWIISC–Geography–Japan–Okinawa–8*

Okinawa, 1993: Yaeju-Dake, the "Big Apple." Japanese soldiers retreating from the Shuri Line defended this strongpoint against 96th Division's attacks in June 1945. *Rodney Earl Walton*

Near Yuza, Okinawa, June 9, 1945: Radioman with an SCR 536 Handy Talkie radio during the battle of the Big Apple. These infantrymen are from the 96th Division's 381st Infantry Regiment. *U.S. Army Signal Corps, SC 209464, U.S. Military History Institute, WWIISC–Geography–Japan–Okinawa–3*

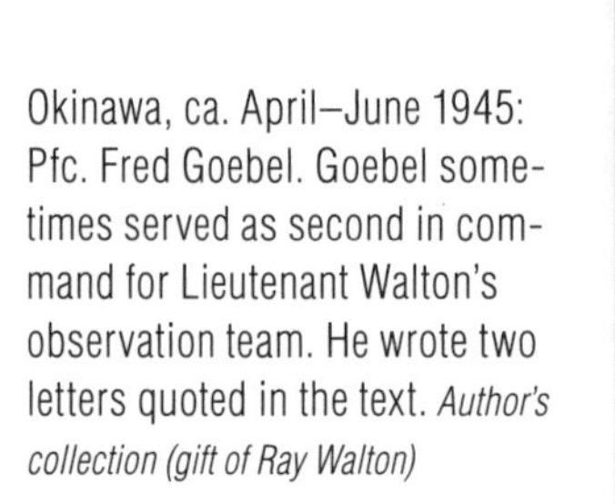

Okinawa, ca. April–June 1945: Pfc. Fred Goebel. Goebel sometimes served as second in command for Lieutenant Walton's observation team. He wrote two letters quoted in the text. *Author's collection (gift of Ray Walton)*

Okinawa, ca. April–June 1945: Capt. Rollin F. Harlow, the commander of B Battery. He was posthumously awarded a Silver Star for his heroism during the closing days of the campaign. *Author's collection (gift of Ray Walton)*

Okinawa, 1945: Okinawan woman carrying a big load while her daughter carries the rest of the family. Although the narrators of this account had only limited contact with civilians during the battle, Okinawa was heavily populated. *Author's collection (gift of Ray Walton)*

Mabuni, Okinawa, 1993: The author (right) and his father (Lieutenant Walton) near a monument to Lt. Gen. Mitsuru Ushijima. The monument stands a few paces from the site of the Japanese commander's suicide. *Author's collection*

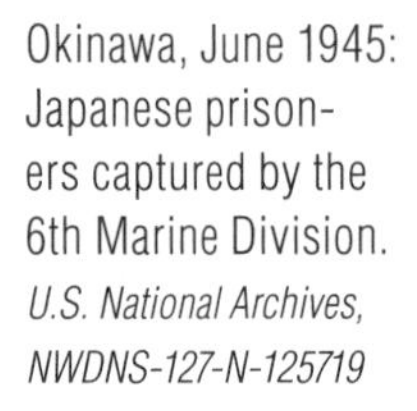

Okinawa, June 1945: Japanese prisoners captured by the 6th Marine Division. *U.S. National Archives, NWDNS-127-N-125719*

Okinawa, 1945: Cub (L-4) aircraft used for aerial artillery observation and transport. An artillery liaison pilot (right) prepares to transport Chaplain Patrick. *Author's collection (gift of Ray Walton)*

CHAPTER 8

THE BATTLE ENDS

The tempo of the U.S. offensive on Okinawa picked up in early June. During the period June 8–11 the roads started to dry.[1] Yaeju-Dake, the "Big Apple," became one of the 96th Division's major objectives. Located in the southern part of the island, the steep cliffs of the escarpment stood as a barrier against the invaders coming from the north. The attack began on June 10. The conquest of this jagged formation of exposed rock cliff became a symbol of the battle of Okinawa. The painter David Pentland, for example, depicted the 381st's infantry accompanied by flame-throwing tanks in the midst of their assault on the Big Apple. Artillery supported the attack. "I remember when we were battling for the Big Apple we were able to spot some machine-gun nests and were able to bring field artillery" to bear on them, recalled Lieutenant Walton.[2]

By the time the attack on the Big Apple began, the Deadeyes had been engaged for two and a half months, and people and equipment alike showed the strain. Capt. Charles Sheahan remembered little about this phase of the campaign. For the exhausted veterans, Okinawa had simply become one geographic obstacle after another. "At times," Sheahan said, "you didn't know the name of the hill you were on. . . . We did the job as we went along. One escarpment after another. . . . And we just kept moving ahead, adjusting artillery as we moved forward. . . . It was one ridge after another." Pfc. Charles Moynihan remembered that by the time they reached the escarpment in the south, "the guns were so badly worn that they weren't accurate. I think there was only one gun out of twelve that they could use with any kind of accuracy."[3]

Private Moynihan began to suffer physical problems about the time the 2nd Battalion reached the ridge. "My hearing got bad, I couldn't hear the radio," he recalled. "I got fungus in my ears and I couldn't hear and so I was relieved. That was down just before we got to the end of the escarpment. And that was just before Buckner got killed." Moynihan returned to the battery. "I could have gone right into the hospital. But I decided to go back to the battery and . . . have our own battery doctor take care of it." After that, he said, "the only thing I had to do was guard duty."[4]

The infantrymen suffered as well. On June 14 Pfc. Curt Sprecher of G Company was wounded when a Japanese round hit the tank he was supporting, and multiple fragments ricocheted into his face. He had not started the Okinawa campaign—having come in as a May 1 replacement—and now he would not be able to finish it. Sgt. William Filter also numbered among the infantry casualties. While on patrol in early June he received a flesh wound in the hip that put him in the hospital for ten days. When he was pressed back into service, his hip was still so swollen that he could not fasten his belt. He was back in action for only four days before he was hit again. On June 15 Filter's twenty-man platoon set up a night defense to block an infiltration route between two large escarpments: the Big Apple on the left and Yuza-Dake on the right. During a night attack, the Japanese cut his communications wire to the company commander, 1st Lt. Frederick Dilg, and about a dozen of his men were wounded. The next morning, June 16, Sergeant Filter went to pick up the gear of one of the wounded men and was shot in the back; the enemy's bullet just missed his spinal cord. He spent months in hospitals before his eventual discharge from Walter Reed Army Hospital.[5]

By June 14 the 381st had gained the heights of the Big Apple, but they still had to fight back counterattacks staged by Japanese soldiers who emerged from their caves after dark. Although the artillerymen were as worn out as their howitzers, the forces of Lt. Gen. Simon Bolivar Buckner Jr. had driven the Japanese toward the tip of the island. As late as May 1945, Emperor Hirohito had clung fast to his belief that American forces would be defeated on Okinawa. By June, however, all hope of that was lost. Historian and Hirohito biographer Herbert P. Bix fixed June 8, 1945, as the turning point. A key adviser drafted a plan asking the emperor to "think seriously about peace." On June 22, 1945, Hirohito told his war council to start taking steps toward ending the war.[6]

The dwindling fortunes of the Japanese did not bring an end to American casualties, which remained high. Artillery observers were no exception. "Of course, most of the casualties were in the FO sections,"

Moynihan said. "They were the most dangerous." Moynihan recalled only one liaison team death in his section, and he considered that one exceptional.[7] Although the battalion commander's headquarters and the artillery liaison officer's position were typically only three hundred feet behind the front line, this small distance made a substantial difference in the degree of exposure to enemy fire on Okinawa.

Walton's five-man observation team suffered thirteen casualties during the course of the Okinawa campaign as new men kept rotating in. Pfc. Fred Goebel was wounded on May 21, 1945. Private Goebel had been awarded a Bronze Star for his heroic action on Kakazu Ridge on April 10 and had been assigned as head of section for Lieutenant Walton's FO team—the highest-ranking position for an enlisted man on the team.[8]

Goebel was evacuated to Guam and soon afterward wrote Walton ("Junior") a letter that reveals much about social relations among the forward observer team members.

June 8, 1945

Dear Junior:

No doubt you are surprised to hear from me so soon. At the present I am in the Marianas at an Army hospital. You can tell Mac [Lt. Samuel C. MacCleod, another FO] that I acquired that billion dollar wound he was always talking about, and it seems I will eventually end up in the States, I hope. The extent of my injuries were a chest wound and my right arm. The wound in my right arm is the one that has done the most damage. It severed a nerve and I have lost control of my wrist and fingers on that arm. However, they say that it can be fixed up OK.

Would certainly like to have a copy of those pictures we took at our 2nd position. Will you give my address to Brogan [the battalion mail clerk] and have him forward my mail to me. Give my regards to [battery commander] Capt. [Rollin] Harlow and the fellows. I'd certainly appreciate hearing from any of them. Has Smitty [chief of detail who was put in for a field commission but may not have gotten it because he got

> sick and was sent back to the States] made his commission yet? I'd better close for now, and please drop me a line when possible.[9]

Goebel's letter reflects the informality among the members of a field artillery observation team. They had slept together in the same muddy holes and eaten together for weeks. Thus Goebel, although a private, first class, felt comfortable calling two officers by their nicknames. In garrison duty this would have constituted an intolerable breach of military courtesy. The letter further demonstrates the limits of informality. It did not extend to the battery commander, Harlow, who would soon be beyond the reach of Goebel's regards.

Captain Harlow fell in action on June 15, 1945, the only captain in the 361st to be killed on Okinawa. Some artillerymen attributed Harlow's death to the shortage of qualified observers. B Battery had run so short of observers, Moynihan noted, that even enlisted men were being pressed into service. "B Battery had gone all the way down to corporals for forward observers to be in charge of sections because they had lost so many men," Moynihan recalled. "They were pressing everybody that was qualified to become an observer."[10] Even the battery commander was forced to do a stint, as Captain Sheahan recalled. "In fact one of the battery commanders became a forward observer. That is a battery commander of the 361st, and he was killed. . . . Harlow . . . , I don't know if [he] was [filling in] for me but it . . . might have been for me. People were reassigned like that—just temporary. . . . You can be a battery commander and still be a forward observer. You are not tied down to your position."[11] "They were taking anybody they could get to go up there," Private Moynihan recalled. "Because I don't know really how many the firing batteries lost . . . but I would guess that they [B Battery] were the heaviest of all [of the artillery batteries]."[12]

Harlow had come up to the front line to bring replacements to Lt. Al DeCrans. Harlow's job rarely called for him to be on the front lines, but DeCrans was short of men that day. The battery was also short of forward observers.[13] Sgt. Jack Gilday, who was normally a section leader for DeCrans, had been pressed into service in an officer's position as a forward observer for E Company, the unit to the left of F Company, where DeCrans was located. Harlow wanted to evaluate Gilday's performance in an officer's role, so he hung his web belt and holster in a tree near DeCrans' position, stuck his .45-caliber pistol in his pocket (so that he would not be marked as an officer), and went over to talk with Capt. George M. Wilcox, the Georgian who commanded E Company. DeCrans could not see E Company from his location two hundred to four hundred yards away, but he heard the rest of the story from Sergeant Gilday, whose observation point was on a ridge. Wilcox took Harlow to the observation point to show him Company E's next objective. Japanese fire struck Wilcox, but he was not mortally wounded. When Harlow attempted to retrieve Wilcox, a Japanese bullet struck him between the eyes. The dead artilleryman became the posthumous recipient of the Silver Star.[14]

Gilday carried Harlow's body down from the ridge. Harlow weighed about 200 pounds; Gilday was about five feet tall and weighed 140 pounds. Gilday was never the same after that incident. He was sent to the hospital for two days and never again served on a forward observation team.[15] The event was painful but less traumatic for other member's of Harlow's unit. Moynihan recalled that Captain Harlow "was well liked." Capt. Lorne I. Martin, who had served a term as the B Battery commander during training, was transferred back from battalion staff to take command again.[16]

In addition to Captain Harlow, the Japanese defenders killed two lieutenants and nine enlisted men from the 361st Field Artillery Battalion. Three officers and twenty-nine enlisted men were wounded in those last days.[17] Many B Battery artillerymen were patients in the Army hospital on Guam where Private Goebel was recuperating. On June 26, Goebel wrote "Junior" Walton about an impromptu battery reunion.

> Hi ya Junior,
>
> . . . In the past week we have had a gathering of Btry "B" here at the hospital. [Pfc. Denver A.] Robinson [wounded June 12, 1945] is in the same ward as I am. [Cpl. Howard E.] Gatenby who is writing this letter is in the nut ward because there wasn't enough room for him anywhere else. Also saw Muscles

> Brogan and [Pfc. Lonnie] Barnhart. I think Brogan [the battalion mail clerk] is headed back now for the b[at]t[e]ry. . . .
>
> We were certainly sorry to hear about Capt. Harlowe [*sic*], and we could hardly believe it when Barnhart [wounded June 16, 1945] told us. It's too bad about Gilday also but that's war. Well *buddy* I guess this is all for now. Write soon.[18]

The high casualties created friction within some teams of frontline artillerymen. An exchange between 1st Lt. Oliver Thompson and his new head of section serves as an example. Thompson was awarded Bronze Stars for his actions on Leyte and Okinawa as well as a Purple Heart. His observation team from B Battery of the 362nd Field Artillery Battalion, however, had suffered unusual losses, and Thompson's battery commander complained that "he was getting tired of writing condolence letters to the next of kin." Early in the Okinawa campaign Japanese machine-gun and mortar fire had killed some of Thompson's men as they attempted to move forward carrying the heavy artillery radio. Thompson quickly became notorious. "I had gained a reputation among the battery personnel of losing too many men and no one wanted to come forward with me," he admitted. Thompson finally got replacements, but among them was an outspoken corporal with no frontline experience. When the lieutenant's team moved into position on the line of battle, the corporal objected. "I told him to set up the radio," Thompson remembered. "He remarked that I was going to get them all killed." Thompson stood his ground. "A dialog ensued where I told him I didn't start this damn war and I stood just as much chance of being killed as he did." The corporal soon "settled down and did his duty throughout the rest of the war."[19]

Several factors were responsible for the high casualty rate among observation team members. First, the team members were in an exposed situation. If they were in a position to see the enemy, the enemy could likely see them. Second, according to Brig. Gen. Robert G. Gard, the division artillery commander, the team members lacked adequate training. He suggested teaching them "scouting and patrolling—infantry tactics. We've lost a lot of good observers because they didn't know how to take care of themselves in an area where Jap snipers and machine guns are a menace." Gard emphasized that "observer parties must know how to take cover, use camouflage—in fact, do everything the front-line infantryman does. On Leyte we sent out forward observers with infantry patrols to train them in such matters. It has paid big dividends, but that's a little late to start training. It should begin back in the States."[20] Third, the Japanese were skilled at identifying and targeting American officers.[21]

Military personnel were not the only ones still in danger after major combat operations drew to a close; local Japanese civilians suffered as well. Although the men I interviewed for this account had only limited exposure to civilians, the inhabitants constituted a potential issue of concern for frontline artillerymen. Since the observers directed fire at long distances, they might be unable to distinguish combatants from civilians. Many of the earlier World War II Pacific battles were fought on largely uninhabited islands. In other areas, such as New Guinea, the civilian population was sparse and identifiably distinct from the Japanese. On Okinawa, for the first time in the Pacific war American soldiers were fighting on Japanese home territory inhabited by a large number of civilians. The Japanese mobilization of Okinawans made distinguishing civilians from combatants even more difficult. Some civilians were mobilized the day before the invasion. Among those drafted on March 31, 1945, to serve in the Blood and Iron Corps for the Emperor were the 360 young students attending the Okinawa Normal School. Twenty-three-year-old Masahide Ota, future Japanese historian of the battle and future governor of Okinawa, was among them.[22]

The artillerymen recognized the problem. The 361st Field Artillery Battalion made every effort to reduce civilian casualties. For example, on May 25, at 4:55 p.m., the battalion reported to the 381st Infantry Regiment, the unit it was supporting: "Art[iller]y Obs[ervers] from OP [observation post] on conical peak observed 40 civilians moving W. . . . Also 30 more, then 80 more moving E at 7966-A, going to same RJ [road junction] and turning S. Definitely civilians—have babies—clothes, etc. They are NOT wearing White clothing. We are NOT firing art[iller]y on them."[23] For the most part, however, the observers from the 361st had little contact with civilians. Much of the 96th Division's heaviest fighting took place in sparsely populated rural areas. Captain Sheahan recalled, "We saw some civilians . . . but not many. . . . Mostly I never saw civilians."[24]

Civilians nevertheless suffered enormous casualties on Okinawa. Although the exact number of deaths is unknown, estimates are usually around 150,000. The casualties were particularly dramatic in southern Okinawa, and many, sadly, were suicides. Convinced by the Japanese that capture by the Americans would inevitably lead to excruciating torture, many Okinawan civilians chose to jump off the steep cliffs into the sea. Just a few years before his death, eighty-one-year-old Willard Bollinger could still visualize them jumping. They were, as armor lieutenant Bob Green correctly observed, "truly between the proverbial rock and hard place."[25]

Years later, Lieutenant Walton still remembered his sole civilian encounter with horror.

> One of my men found a crying baby. It was lying on a large rock on our path as we advanced with the infantry. One side of his face was eaten off by maggots, or rats, or something. I felt very bad and wondered, "Where is the mother?" Either she had been killed or she couldn't help the baby. I didn't know. I always wonder what happened, and remember that baby. There was nothing we could do for it, so we sent it to the field evacuation hospital. The emaciation was deep into the skin and looked terrible. It was about six to eight months old. That was one of my most sad experiences during World War II.[26]

Civilian casualties seemed likely to continue as the war progressed beyond Okinawa. The Japanese government anticipated mobilizing millions of civilians in the defense of Japan proper.[27]

The Okinawa campaign reached its messy climax deep in southern Okinawa in mid-June. A fortified area called the Medeera pocket marked the final point of Japan's organized resistance on the island. On May 24, when it had become clear that the Shuri Line would not hold, the Japanese high command directed that the "main strength of the Artillery Group will occupy strongholds east of [Medeera] and be prepared to concentrate artillery fire whenever necessary." The concentration of enemy forces had important consequences for the 381st Regiment of the 96th Division because Medeera lay squarely in its path.

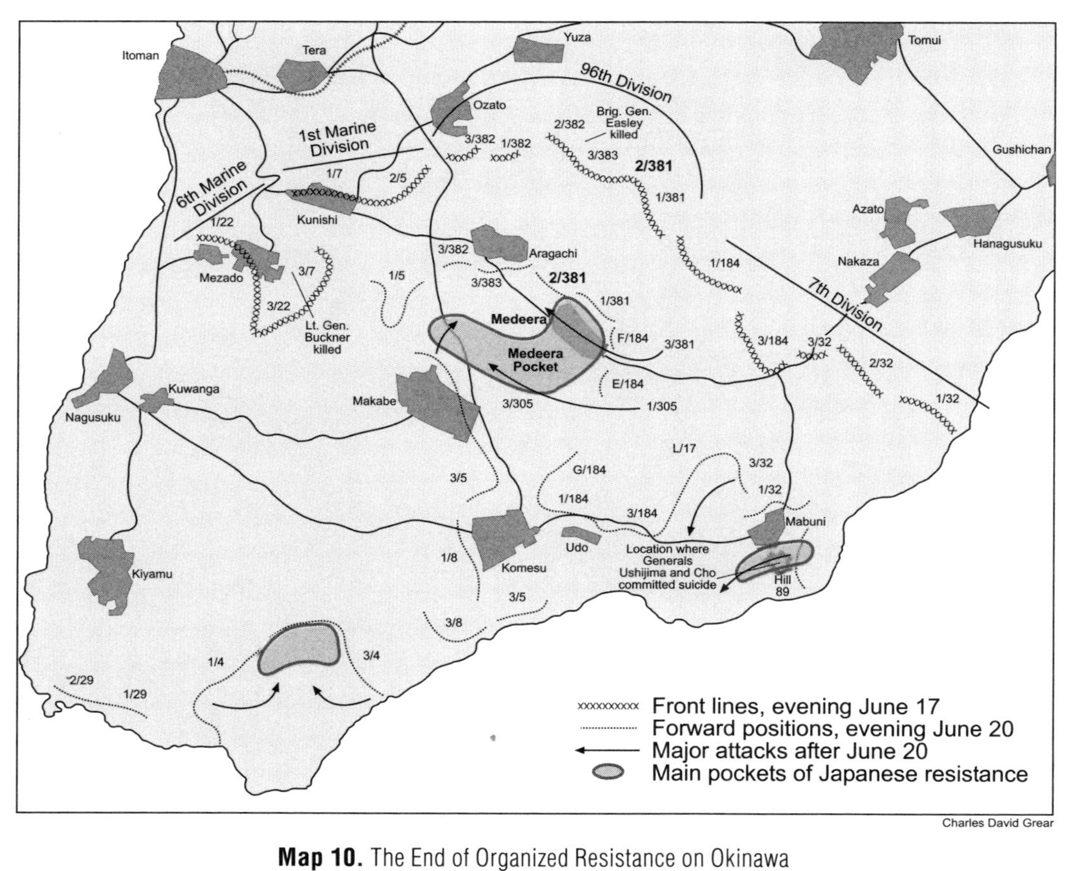

Map 10. The End of Organized Resistance on Okinawa

Adaptation by Charles Grear based on map XLIX in Roy E. Appleman, James M. Burns, Russell A. Gugeler, and John Stevens, *Okinawa: The Last Battle* (Washington, D.C.: Center of Military History, 1948, 1991).

Captain Sheahan remembered the Medeera pocket. His unit had advanced south from the Big Apple, and the Japanese "were all penned in down there."[28]

American artillerymen sought to crush these remnants of resistance. On June 16 the 361st placed several simultaneous concentrations of artillery (time on target) on the nearby towns of Medeera and Makabe. Historians James Belote and William Belote called it "perhaps the most devastating single artillery concentration of the Pacific War." Observers from the 96th Division directed a 264-gun attack on a Japanese troop concentration close to Makabe—two miles from the final Japanese headquarters in southern Okinawa. Shells from all of the howitzers arrived simultaneously and exploded just over the target.[29]

Battle casualties dictated adjustments for some units. Artillery liaison officer Oliver Thompson found his role limited in the closing phase of the battle because the infantry unit he was supporting had been decimated. During the previous month, the first battalion of the 382nd Infantry Regiment had suffered 481 casualties while advancing the relatively short distance between the Shuri Line positions designated Zebra and Oboe. When Col. Edwin T. May, the regimental commander, suggested an attack on the Japanese positions on Oboe Hill, the young infantry battalion commander, Lt. Col. Charles W. Johnson, protested, "With what? I've got nothing left!" Colonel May was adamant. "You know there are no replacements. We have to have Oboe or the third and second [battalions] will be taken in the flank." He ordered Colonel Johnson to "gather up what you have left and attack Oboe." The best the unit could muster was a provisional company—roughly one-fourth the battalion's normal size. Lieutenant Thompson advanced with the infantry because Colonel Johnson "needed every rifle and weapon up on that hill." The battalion took Oboe and then held it against Japanese counterattack. "For the remainder of the campaign," Thompson recalled, "the 1st Battalion fought with 'what they had left.'" The unit's reduced manpower limited its actions to "mainly just cleaning up small pockets." Artillery was not required for infantry objectives of this size, Thompson reported, and thus ended his active combat as an artillery observer.[30]

Although cornered, the Japanese were by no means defeated. They continued to inflict casualties on the American infantrymen. On June 20 the 381st reported "strong pockets of enemy resistance vic[inity] MEDEERA."[31] On the same day, only two days before the "official" end of the campaign, enemy fire killed Maj. Leon Addy of the 2nd Battalion. The major was up front as usual, headed toward a draw in front of G Company to make observations. Men from Sergeant Filter's platoon who visited him in the hospital assured him that Addy had been warned against entering the draw because the Japanese had it zeroed in. Major Addy brushed aside the advice and died.[32]

The casualties continued to mount. On its last rotation to the front lines, DeCrans' team brought along Cpl. Donald Wall from Iowa, who had volunteered for battlefront duty because he had a brother in Captain Bollinger's F Company. The artillery observation position was behind a small hill; the infantry positions were in front of the hill. Bollinger let Wall's brother come back to the artillery observation team's position and talk with his brother briefly before resuming his post on the line of battle. An hour later Corporal Wall watched in anguish as his brother was evacuated from the front with a severe leg wound.[33] The Japanese army refused to recognize their imminent defeat. The American front lines remained potentially lethal. About the same time Corporal Wall's brother was hit, a Japanese bullet passed through Bollinger's pants and shorts as he was inspecting the front lines. Fortunately for the veteran infantry captain, the round left only a six-inch scratch, and he narrowly avoided the sort of disabling wound suffered by his fellow company commander, Capt. Louis Reuter Jr.

Reducing the Medeera pocket posed complex problems for the artillerymen. First, other units—the 7th Division (on the 96th's left) and 1st Marine Division (on the 96th's right)—also took part. By June 20 the 381st Regiment reported to the 96th Division that the "location of friendly troops limited fire to preparatory and defensive missions."[34] Further, operating in close proximity intensified the interservice rivalry between the Army and Marines. Lieutenant Walton remembered that "as we got down there, the Marines started running wild. . . . [They] would take a hill and then run on and not clean it out. And suddenly we had them spotted all over, and we couldn't direct fire because we never knew which hill the Marines were on, and also our own troops were moving quite rapidly, and we were trying to support them, but the situation was very difficult."[35] Thompson, who never met Walton and served in a different artillery battalion, was even more outspoken in his criticism: "When the Marines came into the line in the middle of May, they made derisive remarks about the Army. . . . They did no better than the Army personnel. Their tactics were those of the Civil War. Charge them, then reverse direction and fight back over the same ground."[36]

Second, many artillery pieces were worn out. They could still be used against large, long-distance targets but not for targets requiring precision. Lieutenant DeCrans' last assignment in support of the infantry was with F Company in southern Okinawa. The infantry was in the process of driving the Japanese out of their holes with flamethrowers, and the artillery was of very limited use.[37]

Artillery undertook a new role late in the campaign. At least one of the guns distributed propaganda. SSgt. Karel Knutson helped load leaflets into the shells that Fred Lockman, a section chief for one of the howitzers in B Battery, fired into the Japanese lines. One type of leaflet had a clock face on it with the hands pointed to eleven o'clock, signaling that time was running out for the Japanese.[38]

With the campaign winding down and with artillery fire of little use in close combat, some B Battery gunners turned themselves into infantryman. Thus, toward the very end of the Okinawa campaign, Knutson left his artillery piece behind and joined some other men in search of Japanese soldiers. On the edge of a cane field the group saw three cliff caves with entrances about thirty or forty feet apart, but they did not know which cave the Japanese were in. Knutson spotted flies in front of one entrance and figured the Japanese must be there. They threw a satchel charge in that cave opening. As it turned out, a tunnel connected all three caves, and the charge blew the bodies of some Japanese soldiers out of the cave entrance of another one of the three caves. The Americans found a flag with a rising sun on one of the Japanese bodies. The Americans flipped a coin for possession of the flag; Knutson won. He kept the flag throughout his life. Fifty years later he was still proud of his trophy.[39]

Like the artillerymen, infantrymen found their use of weapons restricted. The danger of friendly fire persisted. Friendly forces made two attacks parallel to the front line of the 2nd Battalion of the 381st Regiment.

The 7th Division made one. On June 21 the 3rd Battalion of the 381st attacked from the west through the town of Medeera.[40] With the use of rifles and machine guns circumscribed, the infantrymen of the 2nd Battalion could only hope that no fleeing Japanese rushed their position. The 2nd Battalion became the anvil against which Japanese troops fleeing the pocket were to be hammered.

As the 96th Division pursued the Japanese Army, some high-level American officers suffered casualties. On June 5, one of the 96th's regimental commanders, Colonel May, was killed near the town of Iwa (now called Kochinda). Brig. Gen. Claude Easley, the assistant commander of the 96th Division, was killed while visiting the front lines south of the Big Apple on June 19. At the time of his death, Easley was pointing out a Japanese machine gun that had wounded his aide.[41] Easley was the last American general to die in combat in World War II.

One critic viewed Easley's death as oddly appropriate. "He was too damn brave, too gung-ho. We were all going to get killed following him," insisted Charles R. "Dick" Thom, a plans and operations officer in the 381st Infantry Regiment on Okinawa. Thom blamed Easley for the heavy American losses on Kakazu Ridge. He gave his views to a high-level panel conducting a postmortem of the tactics and strategy on Okinawa and is quoted as saying, "I did all I could to hang the vast number of casualties on Easley." Pfc. Donald Dencker also criticized Easley for being overenthusiastic and called him foolish "for a general" because he was acting like an infantryman in the incident that got him killed. Lt. Bob Green, a tank officer attached to the 96th Division, recalled that "many of the infantrymen . . . thought it very likely that [General Easley] was going to get them all killed with his plans and hated to see him coming."[42]

In contrast, most of the Deadeyes admired General Easley, even years after his death. Lt. Donald Burrill gave Easley credit for turning the 96th Division into fine marksmen. As a younger man, "Spec" Easley had competed in marksmanship at the world-class level. Capt. Lauren Soth called Easley "among the Army's finest rifle shots." During training, Easley had passionately emphasized rifle instruction. His efforts bore fruit.[43]

Easley's approachability made him popular among the frontline troops. He often appeared at the front sniping with the sharpshooters. Lieutenant Walton reported that General Easley "was really the frontline commander. . . . Every evening he would come about 4 p.m. and make sure all the companies were in the good defensive position . . . for the night. . . . So we saw him frequently." Captain Sheahan recalled: "I was speaking with Easley not too many hours before he was killed. He used to move around a lot, you know. . . . Easley was always exposing himself to fire. . . . I used to talk to him a lot." He would "talk to everybody," Sheahan insisted.[44]

The dramatic death of General Buckner on June 18, marked Japanese artillery's final major triumph. The Japanese command had concentrated all its remaining artillery in the Medeera area.[45] General Buckner had gone forward to an observation post to monitor an assault by a Marine unit, and the Japanese had marked his approach. Buckner's party had come up in several jeeps. A Japanese artillery officer had correctly guessed that the caravan was escorting a person of special significance—one who would justify the use of one of the few rounds of artillery ammunition still available. According to the account of Captain Shoichiro Ishihara, the Japanese used their last 150-mm howitzer to fire the shot.[46] Pvt. Ken Stinson, a rifleman with the 8th Marine Regiment, heard the boom as the Japanese shell landed on the hilltop regimental observation post to his rear. Soon word filtered down the line that Buckner had been killed—not by shell fragments but by a chunk of coral blasted loose by the artillery round.[47]

The death of Lt. Gen. Mitsuru Ushijima was equally dramatic. During the last days of the campaign, the leadership element of the Japanese army on Okinawa was not in the Medeera pocket but more than a mile away, momentarily protected from American firepower in a cave that overlooked the ocean on the back side of Mabuni Hill (Hill 89). The 7th Division was attacking the hill and had approached within hand-grenade range of the entrance to the cave. At around 3:40 in the morning of June 22, after consuming a sumptuous meal, General Ushijima and his subordinate, Lt. Gen. Usamu Cho, left the cave. On a ledge high above the ocean, each man plunged a knife into his own body.[48] The 7th Division captured Mabuni Hill soon afterward.

General Ushijima died only four days after General Buckner and within five miles of the location where Japanese artillery fire felled the American commander. They had much in common. Both were infantry specialists. Both men had held high-level leadership positions in the military academies of their respective nations (West Point and Zama).[49] Each commanded a large army on Okinawa, and neither survived its carnage.

Following the deaths of the commanders, the campaign rapidly drew to a close and gave way to mop-up operations. Historian Thomas Huber noted that "only elements around the headquarters of the 24th Division at Medeera still fought on—but not for long." On June 22 a battalion of the 7th Division attacked and captured Hill 85 south of Medeera. During the last day of the campaign, Captain Sheahan left the front line and returned to his artillery battery. He remembered that the infantry had "kind of stopped still, and then I was kind of free to go either forward [or back]. . . . I decided to go back. I had a choice. . . . I understand there was quite a mess [at the front] which I didn't get to see, I didn't want to anyway."[50] The frontline "mess" consisted of massive Japanese casualties.

Thus ended the greatest field artillery campaign of the Pacific war. On June 22 at 11:36 a.m. the 361st Field Artillery Battalion reported to the 381st that all forward observers and liaison officers had been recalled from their infantry units except for the liaison officer assigned to the 381st regimental headquarters.[51] If the battle of Okinawa was indeed "the last battle," as Appleman and his coauthors called it, then the battle for the Medeera pocket had been the last fight of the last battle.[52] The Americans declared Okinawa secure on June 22, 1945. Soon after the declaration, "Vinegar Joe" Stilwell took command on Okinawa.

Organized resistance had ended, but the fighting went on. Many Japanese soldiers refused to surrender. Small groups and individual Japanese soldiers evaded U.S. forces and posed a continued threat. The 96th Division turned 180 degrees and retraced its steps to the north while sweeping the area for bypassed or infiltrating Japanese soldiers. Sheahan recalled that "several friends of mine got wounded . . . shot up even back there after it was all over, supposedly all over. I remember one fellow . . . went into a cave there looking for souvenirs. Came out. Got shot in the stomach. I was with him when he died." DeCrans recalled that the Japanese did not act like they had lost.[53]

During the mop-up campaign the Japanese lost 8,975 men to the five American divisions. Of necessity, infantry, not artillery, dominated the mop-up operation, but artillerymen remained in harm's way. The last fatality of the 361st Field Artillery Battalion did not take place until August 17, 1945—more than a week after the second atomic bomb had been dropped.[54]

On July 3, 1945, General Stilwell had conducted an awards ceremony for the 96th Division. The Deadeyes were camped in the center of the island, about three-fourths of the way down from the northern tip. Pilot Wayne Welch flew the general to a level open field. The Cub landed near Lieutenant Burrill. After an extended hospitalization on Saipan, Burrill had returned to his battery in early July 1945. Stilwell made the following entry in his diary for July 3: "One-thirty: flew to 96th Division strip for decoration of about 60 men. Unusual number from North Dakota." The general awarded the recently returned Burrill a Silver Star. First Sgt. "Chief" Robertson stood next to Burrill. He, too, received a Silver Star from Stilwell.[55]

Stilwell next visited some Deadeyes who had not been singled out for exceptional awards. B Battery was one of the units he inspected. The observation team members were still not wearing their rank insignia. The general asked DeCrans his name and rank. DeCrans answered that he was a second lieutenant. Stilwell asked him how long he had been a second lieutenant, and DeCrans responded that he considered himself the ranking second lieutenant in his theater of the Pacific. DeCrans had, in fact, been a second lieutenant for more than two years. Because of transfers and illnesses he had never held any one job long enough to get promoted. Stilwell then turned to Sergeant Gilday, who was standing next to DeCrans, and asked how long he had been a sergeant. Gilday replied that he considered himself the ranking sergeant in the theater. General Stilwell said that he would see what he could do for the two of them. Two days later, both DeCrans and Gilday were promoted. Not everyone was so fortunate. Despite service on both Leyte and Okinawa, two Bronze Stars, and a Purple Heart, 1st Lt. Oliver Thompson was twice refused a promotion to captain. On learning of the rebuff, Thompson replied that he "didn't give a damn. . . . I was going home and so many that I knew were not."[56]

The men of the 96th Division had proved tenacious fighters on Okinawa. Captain Bollinger noted that the 381st Regiment "took two of the highest points of terrain in the southern Okinawan battlefields, completed the conquest of Conical Peak, and took Sugar Hill to enable the Shuri Line to be enveloped, and in the vicinity of [Medeera] cleaned up the last holdout enemy pocket." Private Moynihan felt great pride in his division because "they stood their ground. . . . To my knowledge the Deadeyes never broke rank."[57]

CHAPTER 9

SPECIAL TOPICS

Ground observers were not the only men directing artillery fire on Okinawa. Indeed, throughout World War II, small Piper Cub aircraft carrying artillery observers played an important role in American ground warfare. Ken Wakefield observed that "with the exception of the atomic bomb[–]carrying B-29 Superfortress," a light American artillery spotting plane "could bring greater destructive power to bear on a selected target than any other single aircraft in the Second World War."[1] The Cubs participated in the initial invasion of Okinawa, and within days of the landing, artillery observation "Grasshoppers" were flying out of the recently captured Kadena air base. The first American aircraft to land on the captured Okinawa airfields was a Marine observation aircraft on April 2, 1945.[2] In all, the Grasshoppers flew 3,486 missions on Okinawa doing everything from artillery spotting to photoreconnaissance and medical evacuation.[3] The 361st had its own artillery spotter pilots, including Lt. Wayne K. Welch and Capt. John L. Briggs. Both Capt. Charles Sheahan and Capt. Willard Bollinger noted the presence of the small, light aircraft in their interviews, and Sheahan said that they "always did a great job." The role of Lieutenant Welch in the defense of the Tombstone Ridge line is discussed in chapter 6.[4]

Col. Hiromichi Yahara admitted to his interrogators that these small planes inspired terror in the defenders on Okinawa, who "learned quickly that the presence of an observation plane overhead usually presaged enemy fire. And, although they appeared to present fine targets, observation planes were tantalizingly hard to hit with small arms. Observation planes were, therefore, treated with great respect, all movement being kept to an absolute minimum while these planes were overhead."[5]

Historian Thomas Huber noted that the Cubs were also important in setting up counter-battery fire (barrages against Japanese artillery).

"When the Americans got a bearing [on Japanese artillery], they would blanket the area with fire or else send in a cub reconnaissance plane to pinpoint the offending gun and try for a direct hit. The Japanese tried to ward off the cubs with antiaircraft fire or obstruct their observation with smoke, but not always successfully."[6]

Despite the occasional vulnerability of Japanese artillery to attack from the air, the Japanese army on Okinawa had a high opinion of its own gunnery. Colonel Yahara, the highest-ranking Japanese staff officer captured as a result of the battle, took great pride in "the considerable artillery strength" concentrated in Japanese hands on Okinawa and even believed that it wielded "greater offensive power than infantry."[7] It managed to disturb the sleep of Lt. Gen. Simon Bolivar Buckner Jr., whose headquarters was well behind the front lines. On April 26 he wrote to his wife:

> For several mornings, a long range Jap gun has been dropping shells near our C.P. [command post] between 3:45 and 4:00. We can hear the whine of the projectile on its way toward us, followed by the "crump" of the explosion and whistle of fragments and several seconds later by the distant report of the gun. So far he hasn't hit anything or anybody. We are sound-ranging on this battery and have him about located, so our counter-battery will probably get him before long. Apparently he is firing from the mouth of a cave.[8]

Indeed, wherever it was located, Japanese artillery fire could be devastatingly effective. One day as Lt. Al DeCrans' observation team walked south toward the front line, he noted some engineers and Seabees repairing a bridge with a Caterpillar bulldozer. A Japanese shell, probably a 150 mm, hit the Caterpillar and killed four men.[9]

The Japanese did not have formal artillery forward observers. They preferred instead to take advantage of their thorough knowledge of Okinawa to use prearranged fire that was not adjusted on the spot. "If they thought they knew your position, they would shell you all night," remembered Lt. Oliver Thompson. "I experienced such a bombardment more than once." At least one Okinawa veteran disagreed about Japanese forward observers. Donald Dencker noted that "the Japanese had a reinforced concrete bunker at the SW base of Tom Hill close to Shuri. Tom Hill has the highest elevation on the Shuri Line. Artillery observers would observe from the top of Tom Hill and telephone down target information for fire direction from the bunker. Tom Hill remained under Japanese control until they abandoned their underground Shuri Headquarters."[10]

Sergeant Knutson observed that the Japanese artillery was "zeroed in" all over the island; that is, the gunners had already test-fired their weapons on pre-positioned targets.[11] Artillery telephone operator Roman Klimkowicz recalled that the Japanese regularly fired on road junctions, particularly at night. This appeared to be prearranged fire intended to catch American vehicles using the roads after dark. Klimkowicz recalled one road junction not far from B Battery's position during the middle phase of the Okinawa campaign. During the day, an American aid station was located near the junction. The station was just a shack that gave the medics and the wounded some shade from the sun. In order to avoid exposing the position of their guns, Japanese artillery rarely fired on the road junction during daylight hours, but they regularly shelled it at night, blowing apart the aid station each time. The next morning the American medical personnel would piece together the remnants of the shack and go about their business. In that sense, Klimkowicz observed, Japanese artillery fire rarely disrupted American artillery operations to the extent that it might have.[12]

Some Japanese officers may have served as ad hoc forward observers. Capt. Lauren Soth, an artillery officer, reported hearing that several Japanese officers carrying radios had been killed or captured behind the American lines and that the Americans "believed that they were either adjusting artillery fires or reporting hostile locations to their artillery." Japanese artillery communications appeared rather poor to the American artillerymen opposing them. Soth observed that "the Jap wire lines have been shot out repeatedly, judging by the slowness with which enemy artillery brings fire on our troops, vehicles, and installations under their observation."[13]

Although the Japanese did not make much use of artillery observers, they did have mortar observers. Sometimes American observers found themselves pressed into action as infantrymen to suppress these spotters. Lt. Donald Burrill recalled one such incident on about April 10. He had relieved another observation team, and as he took up his position he "spotted a helmet come up out of a spider hole and look all the way around and go back down. Pretty soon a mortar shell would come flying our way." Burrill took a shot at the Japanese soldier with an M-1 rifle, but he was using an unfamiliar weapon and the shot went low. Burrill adjusted the sight, fired again, and hit the helmet. "I figure I got him," he recalled. "[S]omething that a forward observer always wants to do is rub out the enemy's forward observer. I'm sure I got one that day." Lt. Bob Green likewise shot a Japanese forward observer. It happened on Hacksaw Ridge, and Green also used an M-1.[14]

Japanese artillery suffered from deficiencies other than lack of formal observers. Pfc. Charles Moynihan noted some of them. The Japanese, for example, failed to adjust their artillery. "For some reason they would bracket you with artillery: . . . one in front and one in back," he recalled. "They never threw the third one because we would have been dead a long time ago." In Moynihan's opinion, their failure to fire that third shot, which would have landed right in the middle of the Americans, substantially reduced the effectiveness of Japanese artillery on the GIs and their supplies. At least one other Okinawa veteran, Donald Dencker, maintained that Japanese artillery did fire three (or more) shots at a time. Sgt. Karel Knutson noted another deficiency: duds. Many shells landed in the right spot only to sink into the ground without exploding. There were close calls, but Knutson was "very lucky."[15] The Japanese artillery also proved inefficient in failing to mass multiple guns on specific targets. The Japanese generally could not muster more than a four-gun battery on a single target. Col. Bernard S. Waterman, a high-level American artillery officer, attributed the Japanese failure to use their substantial artillery as effectively as they might have, despite "evidence of careful and intelligent preparation," to poor communications, poor technique, or perhaps both. Prisoner-of-war interrogations revealed that American artillery had disrupted Japanese communications, but Waterman wondered at the failure of the Japanese "to install a network of buried wire communications" that would have been immune to artillery fire.[16] Massed gunfire as the Americans used it was very effective; at one point, 264 guns were firing on a single target. Lieutenant DeCrans believed that the ability to concentrate fire was the major strength of U.S. artillery on Okinawa.[17]

Aside from the heavy fire he received on Kakazu Ridge, Captain Sheahan was not impressed with Japanese artillery. "They never used their artillery as efficiently as we did," he said. "They never had concentrations of artillery. They had single rounds here and there." On the other hand, Sheahan praised the American artillery, calling it "tremendous." Infantry sergeant William Filter described Japanese artillery as having the island well covered with "sporadic" fire but "not as good as ours."[18]

The success of American artillery on Okinawa did not come without mistakes. Field artillery is subject to inaccuracy, and American artillery fire did occasionally explode on or near friendly forces. A European theater observer (Lieutenant Major) reported that "even if no one made a mistake, the normal dispersion [of 155-mm medium artillery] might cause a shell to fall 50 yards long or short."[19] The forward observers I interviewed typically referred to such incidents as "short rounds," but in contemporary military language they are called "friendly fire" or "blue-on-blue" incidents. Although relatively rare, these incidents are nonetheless noteworthy.

Mistakes in laying in (setting up) the howitzers were among the primary reasons for short rounds. Lieutenant Thompson recalled one short-round incident involving worn-out gun tubes. He had notified the 362nd Field Artillery Battalion's fire direction center "that some artillery piece was firing short (but in front of us). When it went overhead it had a loud peculiar sound like it was 'tumbling' thru the air." He later heard from Maj. Miles W. Haberly, the 362nd Field Artillery Battalion's executive officer, that the 362nd was not responsible. "They found the culprit in another battalion. It seemed that a piece had lost its 'rifling' and the projectile was *tumbling* thru the air."[20]

Sometimes the observer miscalculated the coordinates, and sometimes the battery (or tank or self-propelled cannon) failed to comply with the observer's instructions; and, like the Japanese, Americans occasionally fired faulty ammunition. A secondary source of inaccuracy lay in the observer's inability to see the target, something that was particularly problematic against a dug-in enemy. Thus, for example, Japanese forces frequently placed strong points on the reverse slopes, knowing that the American observers could not see the back side of the hill.

Any of these errors might cause a short-round incident. Dencker, estimated that friendly fire caused 8–10 percent of the American casualties on Leyte and Okinawa. Dencker blamed these incidents on "a short round or faulty identification" and concluded that "becoming a 'friendly fire' casualty was always a risk endured by members of a front-line infantry company."[21] Such situations were not unique to Okinawa.[22] Friendly fire was an enormous problem during World War I; indeed, artillery errors were probably less common during World War II. Nonetheless, they did occur. Just seven months before the landings on Okinawa, the U.S. 85th Division in Italy was bombarded by American artillery on four separate instances over the course of four days.[23]

Captain Sheahan considered short rounds the biggest failure of U.S. artillery on Okinawa and a problem in every operation: "Friendly fire is an awful thing. . . . It happens every day. . . .Yah, that short round was a problem." Overall, Sheahan thought that the U.S. artillery support "was always pretty good," but he admitted witnessing some artillery mistakes.[24]

Typically it was the infantry, the forces closest to the enemy, who fell victim to American friendly fire. A short round inflicted casualties when it landed in one of F Company's machine-gun sections and resulted in one death. Captain Bollinger never found out who fired the round, and that might not even have been relevant. He pointed out that human error was not the only possible source of friendly fire. It might have been due to faulty manufacturing. He doubted that the incident was even reported. "It happened" he explained, but added that it was demoralizing—bad enough to get shot at by the enemy without being hit with your own artillery.[25]

Marine rifleman Jim Boan recounted the following short-round incident against infantry in southern Okinawa in his memoir: "Shells were exploding about 150 feet away. 'Chronis, get on your radio and tell that cockeyed artillery captain to lift his sights.' It was the [reconnaissance company] commander giving orders, and he was cussing mad. But the artillery spotter on the hill had seen the error and already made changes. It was too late for George Taylor. A chunk of smoking shrapnel had burned through his shoe and buried itself in his foot."[26]

Captain Sheahan recalled another such incident that took place as he was attempting to direct fire for an artillery unit that was not part of the 96th Division. It was mid-campaign, somewhere in the vicinity of Conical Hill. The men who were hit were from the 1st Battalion. On new targets Sheahan typically fired smoke rounds before firing high explosives so that he could see precisely where the rounds were landing. If the rounds went astray, smoke rounds caused less damage than high-explosive ammunition. Sheahan asked for smoke first, but the firing battery refused, claiming that it had already marked the position. "I said, 'Well throw a smoke out and I'll bring it in.' And they said, 'No, no, we know where it is.' It wasn't our outfit but they . . . said they knew what they were doing. They're sure. . . . So they goofed up [and] . . . sent a phosphorus shell. And it landed right among a bunch of our own men."[27] The overconfident unit had fired white phosphorus, a chemical that would cling to the skin of any human in the path of its explosion and burn right through any clothing. Attempting to wash it off with water only caused it to burn more. The phosphorus round landed in an emplacement occupied by U.S. infantry troops. Sheahan recalled that "there were about eight men in a group. They were a little apart from each other. I remember very distinctly. . . . There was a lot of pain down there I am sure." Sheahan believed that some of the Americans must have been killed.[28]

Like infantrymen, military engineers also suffered from friendly fire. These soldiers often needed to stand upright to perform construction tasks. Thus, even though engineers often worked behind the front lines, they were more exposed to airborne shrapnel than frontline soldiers, who could hug the ground. Captain Sheahan witnessed one fatal result of a short round against an engineer lieutenant. "I remember the short round . . . on one of those ridges I was talking about when a round come in, blew him to pieces, he had taken his knapsack off, it was World War I [equipment] and his name was King, I remember, so I picked it up and I had it in Korea even."[29]

On the afternoon of May 31 a group of military engineers attached to the 381st (for the purpose of keeping the regimental supply road open) likewise felt the sting of friendly fire. A short round killed one officer and five enlisted men and wounded another two enlisted men. The incident prompted an angry comment from Lt. Col. John M. Williams, the 381st's second in command, in his daily report. This was, he steamed,

> the fourth repetition of faulty Art[iller]y firing to have caused casualties in our sector during this campaign. Total casualties caused by these "accidents" now exceed 50. The ratio of replacements which now constitute our Co[mpany]s is generally known. No amount of energetic command indoctrination can possibly cause these continued failures to make sense to these men. Greater coordination must be obtained by higher headquarters than can be accomplished through RGT [Regimental] channels if this natural lack of confidence in our Art[iller]y firing, or our ability to prevent these instances, is to be overcome.[30]

Sheahan judged this incident, which he did not witness, to be gross negligence. Engineers would have been very far from the front line if they were working on a road. Any round that hit them would have been "way short."[31]

Friendly fire even struck American artillery observers. Captain Sheahan observed a short-round incident that may have been fatal to one forward artillery observer: "I was on top of a ridge . . . where I had slept the night and another forward observer came in . . . so I moved down a little bit, dug another hole. While I was digging in the hole when a round comes in—zoom—right where my head had been and killed a bunch of people on top of that hill about fifty or sixty feet from me."[32]

Armor lieutenant Green reported an almost identical incident that occurred at an observation post atop Hacksaw Ridge where numerous forward observers for air, naval gunfire, field artillery, and armor forces were supporting the 381st Regiment.

> The very ground we were sitting on rocked and rolled, and my ears rang and hurt from the sudden roar of a great explosion very close to us. . . . Looking down, I saw a large, quivering, dark red mound of something that looked like raw liver in my lap. . . . At least two 155-mm rounds from our own "Long Tom" artillery that were located several miles behind us had for some reason fallen short and failed to clear the ridge we were on. They had struck squarely on the top of the Maeda Escarpment OP among the crowd of people there. One had exploded immediately

> behind the trench where I had just sunk down to read the news sheet. Two or three observers in the trench with me were sheared off at the waist by flying shards of steel shrapnel. . . . Seven or eight more men were wiped out or wounded when the other shells landed in their midst.[33]

Sergeant Filter of G Company also saw forward observers fall victim to friendly rounds. In the Yonabaru area, two forward observers were directing fire from mobile 105-mm guns (formed into a unit called a "cannon company"). One of the rounds fell short and killed both observers.[34] Captain Sheahan was almost a victim of friendly fire himself: "Well, short rounds, . . . another time I was on top of a ridge and I was just going along and zoom it comes right down almost over my head and killed a bunch of guys, hit a guy in the head . . . pretty gruesome."[35]

Short rounds could cause artillerymen's tempers to flare. On one occasion Lieutenant DeCrans observed repeated incidents of friendly fire. He lay in an advanced position; his chief subordinate, Sgt. Jack Gilday, was behind him with a radio. Navy airplanes dropped bombs about five hundred yards behind DeCrans, and naval gunfire from both sides of the island started landing short rounds behind him. At the same time the Japanese artillery opened up to the front of DeCrans. When Gilday reported that an American dud shell had landed about fifteen yards behind him, DeCrans told Gilday to call Lt. Col. Avery Masters (the artillery battalion commander) and tell him to stop every gun on the island. The firing stopped, and Lieutenant Colonel Masters ordered DeCrans and Gilday to dig up the dud so he could identify the firing unit. Masters' order infuriated DeCrans, who was operating in an exposed position, and he instructed Gilday to tell the colonel that "if he wants that blankety blank shell dug up, . . . to dig it up himself." Gilday did as he was told and transmitted the message over the airwaves. Many heard it, including DeCrans' superior officers. DeCrans expected to be demoted to private, but nothing happened. Eventually it was determined that an American tank had fired the short round. Gilday joked that he had been subjected to high-explosive fire from the right, the left, the front, above, and behind, and he expected it next from below.[36]

Observers who escaped physical injury from friendly fire might yet face it from fellow Americans. Forward observers were subject to accusations, ostracism, and sometimes even physical assault from their fellow GIs. Aside from the infantrymen's obvious instinct of self-preservation, they lived in a constant state of tension with their nerves perpetually on edge. The observers were outsiders—artillerymen in an infantry unit—and some were relatively green lieutenants supporting veteran combat units. Unlike the infantrymen, the artillerymen generally had to spend only three days at a time on the front lines. Thus the observer's situation was rife with possibilities for animosity.[37]

American observers encountered hostile infantry reactions to friendly fire in many theaters of World War II. Artilleryman K. P. Jones related an incident (in the U.S. Third Army) that occurred in Europe late in 1944 when American forward observers called in artillery fire on the infantry company Jones was accompanying. "We called for cease-fire and ducked, but the rounds killed some our own infantry and blew up a bunch of rations," Jones wrote. "One fellow ran wild and was tackled by another."[38] Historian James Russell Major, formerly a forward observer in the European theater, experienced a close brush with friendly fire when the battery for which he was observing fired a short round that "landed about 25 yards behind me in the midst of the infantry. It was a miracle that no one was hurt." Despite the absence of injuries, the infantry seethed. "Artillery observers who fired short were not popular with the infantry, and I expect that I lost some of the good will I had earned by saving their necks the night before," Major recalled.[39] Edwin Westrate described some of the tension between the two branches in North Africa. One disgruntled American infantry company commander cynically commented to his attached forward observer that "he hoped our howitzers wouldn't knock off too many of his men."[40]

Similar incidents took place on Okinawa. Lt. Donald Burrill had a brush with blue-on-blue just two or three days before the April 19 battle for which he received a Silver Star. One of the rounds he requested fell about one hundred yards behind him, exploded in a tree, and hurt some infantrymen. "The infantry were about ready to hang me till [B Company commander Capt.] John Byers jumped in and said, 'You dumb son-of-a-bitch, get back to work. It wasn't this man's fault it happened.'"

Burrill reported the short round and found out that the gunnery corporal back at the battery had been one hundred millimeters off on the elevation setting. "He was the cause of it all. So that absolved me," Burrill said.[41] Burrill's experience demonstrates two facts of life for observers: they could expect to be blamed for short rounds regardless of the cause or origin, and their safety from irate infantrymen could depend on the willingness of a respected infantry officer to stick up for them. James Russell Major was fortunate enough to have another officer there to defend him when he was similarly accused in Europe. The officer had witnessed Major's actions in Mairy, for which he was awarded a Silver Star. "I was in a village when one of our 240mm howitzer shells landed on a barn killing 10 or 12 of our soldiers," Major wrote in his memoir. "I had nothing to do with it but was blamed by some. Fortunately, the lieutenant who had pointed out the tanks at Mairy was there and stilled the troubled waters."[42]

Sometimes the infantrymen's reaction to friendly fire was less hostility than anguish. Lieutenant Walton related that his worst day on Okinawa resulted from a short round fired by one of four tanks being used in an indirect fire role. Walton was acting as the observer for those tanks.

> One day I was shooting a battery of tanks . . . I guess 75-mm guns. . . . Suddenly one of my four rounds fell short and fell in the 96th Division Infantry lines to our left. And I immediately called for "cease fire" because of our short round. And suddenly one of our infantry people said, "May I borrow your field glasses?" And he looked over and then he looked back at me and said, "You've just wounded my best friend!"
>
> And boy, that was probably one of the worst experiences I had in the whole campaign! Now I knew my directions must have been correct or the three rounds wouldn't have gone to the target. That one round that was short could be faulty manufacture or it could have been faulty adjustment of the guns or any other thing, but it didn't help to tell this other GI what happened. He was very sad and was very unhappy with me.[43]

American commanders sometimes had no choice but to expose U.S. troops to friendly artillery fire. Although I uncovered no such incidents in my research of the Okinawa campaign, they have been documented elsewhere. Anecdotally such incidents tend to be associated with defensive struggles. (This would include calling artillery fire down on one's own position to keep it from being overrun by the enemy.) Marine artillery commander Brig. Gen. Pedro del Valle (later to command the 1st Marine Division on Okinawa) described one such incident on Guadalcanal in 1942. Artillery support was called in dangerously close to support Marines defending Bloody Ridge (overlooking the critical Henderson airfield) from a Japanese night attack. "The danger of hitting our own troops was disregarded in the effort to place a wall of fire between our men and the enemy," del Valle wrote. "Two shells did indeed burst in the tree tops over the Division Command Post, causing slight injuries to six men, but no protest resulted as everyone understood the seriousness of the situation and inestimable value of the artillery on that strenuous night."[44]

American infantrymen deliberately exposed themselves to friendly fire on some occasions. Despite the risk of injury, some attacking American GIs preferred to close in behind moving American artillery barrages, hoping to reduce the enemy's time to react to the attack once the barrage had passed over their defensive positions. Forward observer Lt. Richard D. Bush observed this during the Tunisian campaign in World War II. Despite being told to stay two hundred yards behind the artillery barrage being fired to their front, American infantry closed to within one hundred yards and then fifty yards. After the successful attack on German positions, the infantry battalion commander explained: "You know, we only had two casualties in taking that hill, and they were slight wounds from our own artillery fire. Hell, the boys don't mind that—they think it's kind of friendly."[45]

Despite the persistence of short-round problems, infantry complaints about friendly fire were fairly rare. The infantrymen recognized that the observers, being on the front lines themselves, were just as likely to be injured as other members of the unit.[46] Furthermore, infantrymen were often very thankful for the fire support that the forward observers provided. And they maintained a fatalistic view of short rounds. One Okinawa veteran wrote, "Such is what happens in combat."[47]

The Deadeyes were subject to friendly fire from sources other than their own artillery. On Okinawa, 96th Division troops frequently blamed adjoining Army or Marine divisions for short rounds. Lieutenant Colonel Williams, the second-in-command of the 381st Infantry Regiment, wrote a blistering report to 96th Division headquarters as the unit was being moved into rest camp after duty on Hacksaw Ridge, describing "an urgent need . . . for more proficient tactical coordination between units higher than adjoining regiments. This criticism is particularly directed at slipshod and inadequate liaison which permits the artillery of adjacent units to inflict heavy casualties on our own troops when permission to fire has either been unsolicited or denied. Such mishaps occasionally are understandable accidents, but repetition is intolerable. The infantrymen who survive do not feel satisfied all is being done that should be done when these things reoccur with frequency."[48]

American air and naval forces also mistakenly delivered fire on friendly forces. On Leyte, for example, Captain Sheahan had climbed a muddy hill expecting to find friendly troops at the top but had instead found the consequences of friendly fire. "Just about everybody was dead up there, a naval shell had landed and blown all kinds of things apart. At least we blame it on the Navy—I suppose it was them." Similarly, a U.S. Navy destroyer shelled elements of the 96th Division shortly after they landed on Leyte (October 20, 1944) and wounded about a dozen men.[49]

A Marine Corsair was reported to have strafed the 2nd Battalion of the 381st on May 26. Marine rifleman Jim Boan recounted an incident in which an American plane killed a Marine laying communication wire. After the battle on Okinawa, Lieutenant Green wrote to his mother that "our own planes made many mistakes." Green described how a U.S. Navy plane killed an infantry battalion commander with .50-caliber machine-gun fire while Green was drinking coffee with him.[50]

Japanese forces took advantage of the friendly fire incidents to create confusion among the American troops. They became expert at simulating American short rounds and were doing so as early as the Guadalcanal campaign.[51] Months before the battle of Okinawa, Lt. Col. Robert C. Gildart warned his fellow artillerymen about such tactics in a professional journal article.

> Japanese trickery . . . caused headaches for the forward observers. The Japanese frequently withheld their artillery and mortar

> fire until friendly artillery began its fire for effect. At this point the Japs would fire close to our line to give the impression that "Artillery is falling short." This happened in many cases even when friendly fire had been adjusted six to eight hundred yards in front of friendly troops. The natural result was for the infantry to call *Cease Firing*. It was only after artillery continued to "fall short" after *Cease Fire* that friendly troops became conscious of this ruse.[52]

Lieutenant Thompson of the 96th Division became aware of the strategy when he complained about short rounds falling close to friendly positions during combat. His artillery battalion's executive officer, Major Haberly, admonished him. "He said the Japs have artillery and mortars, too," Thompson recalled. "He was correct. I was green at the time."[53] Accordingly, it is entirely possible that some alleged friendly fire incidents were in fact Japanese efforts to reduce the amount of American artillery fire on their positions.

Experienced artillery liaison officers could help reduce the friendly fire problem. Captain Sheahan, for example, working at the battalion level, was able to guide the observers supporting smaller units within the battalion. Burrill acknowledged Sheahan's assistance. Sheahan typically handled the more delicate assignments—when fire was called in close to friendly lines.[54]

Close cooperation between the infantry and artillery was essential for avoiding short rounds. Decisions about night defensive fire were particularly delicate. The Japanese typically attacked by night, when American infantrymen were dug in and the threat from American air and naval gunfire decreased. Field artillery still played a role at night, but usually only through preplanned defensive fire. Some American company commanders preferred to keep night defensive fire at some distance from their front line. Under these circumstances, the likelihood of a short round decreased but the American frontline infantry received less protection from enemy assault forces assembling in their immediate front. Other American company commanders preferred preplanned night defensive fire immediately in front of their line. Such fire could be devastating to enemy troops gathering for an attack, but it was not without danger, as Lieutenant Thompson observed. "I got along well with company commanders and platoon leaders. [But] . . . they wanted you to fire close to their front lines not understanding the dispersion patterns of artillery. (No two rounds ever landed in the same hole.) If you fired too close then they would complain about the 'swish' of the shells overhead."[55] Nonetheless, observers tried to comply with the company commander's preference rather than make firing decisions themselves. The company commander was expected to know the precise location of his forward troops and to understand the risk of short rounds.

In the final analysis, numerous conclusions can be drawn about American friendly fire and the Okinawa experience. Friendly fire from field artillery was uncommon. Both Burrill and Walton reported witnessing only one friendly fire incident during their participation in the Okinawa campaign. The relatively junior rank of the observers had surprisingly little impact on how infantry reacted to friendly fire. It is true that the observers were often second lieutenants—the lowest officer rank. It is also true that personnel of this rank in any era have often been viewed as inexperienced "shavetails." On the surface, this would seem to create an explosive situation in friendly fire situations. Veteran infantrymen—tired, on edge, and armed to the teeth—might immediately blame the artilleryman in their midst for harm to their comrades. The nature of the Okinawa campaign, however, may have lessened any such impact for several reasons. The high casualties among infantry personnel resulted in constant turnover. The replacement troops, many of whom were from the high school class of 1944, must have viewed the observers as relatively seasoned. The scarcity of junior infantry officers on the Okinawa front lines may have made the artillery lieutenants seem relatively senior and experienced. The infantrymen were usually happy to have observers present to provide artillery support regardless of rank.

I found no evidence suggesting that friendly fire resulted in a decrease in the number of fire missions requested. Furthermore, no evidence appears to suggest that American friendly fire adversely affected the effectiveness of artillery fire against the Japanese. Company commanders who elected to have preplanned night defensive fire in close to their own front lines obviously believed that American artillery was saving more lives than were being lost as a result of short rounds.

In the event of friendly fire, good relations between the artillerymen and the infantry leadership were of central importance. As the case of Lieutenant Burrill indicates, the physical safety of the observer might depend on the willingness of the local infantry company commander to stick up for him. In the end, though, the infantry badly needed the firepower provided by the artillery and usually recognized that some friendly fire was unavoidable. Nonetheless, the infrequency of short rounds was a criterion by which the infantry informally evaluated the proficiency of the forward observers. As one 96th Division infantryman put it, the Deadeyes learned early (during stateside training in approximately May–June 1944) that "close support could get a little too close."[56]

Controversy surrounds the issue of whether American artillery on Okinawa was successful enough to justify the U.S. Army's heavy reliance on it. The dispute over the quality of Buckner's generalship during the campaign is an integral part of the debate. General Buckner followed standard U.S. Army doctrine when he depended on artillery to defeat the Japanese. On May 2, 1945, the *New York Herald Tribune* quoted Buckner on that point: "We're relying on our tremendous fire power and trying to crush them by weight of weapons."[57]

World War II–era U.S. Army doctrine held that artillery could be decisive in combat. George S. Patton, an armor officer, asserted: "I do not have to tell you who won the war. You know our artillery did." This position continues to receive support in the twenty-first century. Historian Janice E. McKenney asserted that the U.S. Army field artillery reached its zenith in World War II and that the artillery was "a decisive factor in the Allied victory."[58]

Despite Army doctrine, the wisdom of Buckner's reliance on artillery on Okinawa was a matter of controversy even as the battle raged. "Vinegar Joe" Stilwell, the caustic general who took command of the Okinawa operation near the end of the campaign, criticized Buckner sharply for overemphasizing artillery.[59] Likewise, Brig. Gen. O. P. Smith, the deputy chief of staff in Buckner's Tenth Army and a Marine, "thought Buckner much too optimistic about the ability of artillery to batter a breakthrough."[60]

Field artillery is of limited efficacy against caves, bunkers, and pillboxes—the sort of targets that were common on Okinawa. The trajectory (curved flight) of artillery shells is not as accurate as direct fire in such circumstances. The Japanese reverse-slope strategy limited the impact of artillery support close to U.S. lines as well. Colonel Yahara later told his interrogators that Japanese planning for the Okinawa campaign had largely neutralized the potential destructive force of U.S. artillery. Yahara insisted that "the effectiveness of [U.S.] Art[iller]y was countered, successfully to a great extent, by the elaborate system of underground fortifications. Heavy bombardments, such as came before attacks caused relatively low casualties."[61] Some twenty-first-century observers support Yahara's position. Journalist Bill Sloan asserted that "there was overwhelming evidence that artillery was virtually impotent against" defenses like Hacksaw Ridge and that "the heaviest artillery barrage yet seen in the Pacific had failed to dent the Japanese defenses at Kakazu."[62]

Although American artillery—pace Patton—did not turn the tide on Okinawa, it remained a critical part of the offensive. Flamethrowers, explosives, and direct fire (for example, using tank or antitank weapons) might have been more effective against Japanese-held caves and hillside emplacements, but artillery could attack these difficult targets from a greater distance and with some success. On April 23, for example, the regimental operations officer reported that the 361st Field Artillery Battalion had destroyed two pillboxes: "two Japs blown out of each pill box when direct hits were made." On May 28 the 2nd Battalion reported "time delay action fire on hill 8170Gl. FO reports many cave-ins and damage." Both Bollinger and Walton remarked that the April 23 incident was unusual because artillery fire was generally not so accurate.[63]

Like Buckner, many grunts considered artillery effective and admired artillerymen. Sergeant Filter, for example, told the story of four American soldiers who found themselves isolated in front of his company for two nights. Compelled to hide in close proximity to the enemy, they unwillingly served as on-the-spot observers of the effects of U.S. artillery, both 105 mm and 155 mm. When they got back to U.S. lines, they told Sergeant Filter, "We're happy we're not the enemy!" They reported that the Japanese soldiers were stunned and dazed by the shelling, and some had practically walked on top of the four cut-off Americans but had been too disoriented to notice them.[64]

Some Japanese acknowledged the contributions made by American artillery. Capt. Tadashi Kojo, a battalion commander in the 22nd Regiment of the Japanese Imperial Army, was one of them. On Okinawa his unit at various times opposed elements of both the 96th Division and the 7th Division. On his first encounter with U.S. artillery and naval gunfire, Kojo "registered unhappily to himself that American firepower was 'unbelievably' more powerful than anything he had expected . . . on subsequent days . . . it kept him underground until it let up at twilight." Author George Feifer vividly depicted artillery's impact on one of Captain Kojo's soldiers: "In Kojo's battalion, Yoshio Kobayashi perceived the enemy's firepower with typical dread. 'There was no dead angle or safe place anywhere. Bombs and shells came from land, from the sea, from the sky—from every angle. If you were in a valley, trench mortars did the job. One step out of your cave and your fate was in the hands of God. Every inch ahead was a black unknown.' A fellow target put it more directly: 'It's sheer wonder that any foot soldiers managed to live.'"[65] Clearly, U.S. artillery on Okinawa could and did provide suppressive fire that strongly encouraged the defenders to remain under cover during American assaults.

Artillery private Moynihan praised the American gunners: "They did a good job! Those fellows worked around the clock. . . . They had to move a lot of ammunition." (Each shell weighed thirty-five pounds.) Moynihan also gave much credit to the recently developed fire direction centers: "Our fire direction center, if we needed more support, then we could call in the 363 [Field Artillery Battalion] which is a 150 [actually a 155-mm howitzer] outfit, which was a heavy one, larger projectiles. And we could also pull in the other batteries [from the other field artillery battalions of the 96th Division] if needed and vice versa. They were all tied in together. And also . . . the fire direction center . . . could call in a lot of DivArty [Division] Artillery." Moynihan considered the artillery most successful at Kakazu and Needle Rock. In both locations American artillery fired "a very successful moving barrage" ahead of the ground troops.[66]

Captain Sheahan declined to single out any one big success for artillery on Okinawa. He viewed the accumulation of smaller successes as the artillery's best contribution. Sheahan remembered some moments of intense job satisfaction on Okinawa. "The big success is when you adjust something, then fire for effect; . . . that's when a whole battalion pounces on them," he recalled. "There was so many of that I wouldn't know any particular spot. . . . A whole battalion firing for effect was pretty—did a lot of damage." The U.S. artillery had "some terrific concentrations that would blow everything," he recalled.[67]

CHAPTER 10

AFTERMATH

More than 500,000 American servicemen participated in the battle for Okinawa.[1] If the air-sea battle that stretched seven hundred miles from Formosa to the southern tip of Kyushu is included in the count, one million men from all sides (Japanese, American, and British) participated in the campaign. Despite the horror of the battle, however, many American veterans of it shared the opinion of Adm. Raymond A. Spruance that Okinawa was merely "a bloody, hellish prelude to the invasion of Japan."[2] When the bulk of the fighting on Okinawa was completed, American military planners found that they had too many combat personnel on Okinawa. The excess forces had to be redeployed to less exposed positions where they could refit and train for the upcoming invasion of Japan—away from the prying eyes of the occasional Japanese aircraft that penetrated Okinawan airspace.

It was no secret that the Americans were planning to invade Japan. During the course of the battle of Okinawa, Capt. Lauren Soth, an artillery officer, wrote an article for publication in the August 1945 edition of the *Field Artillery Journal*. He explained, "I've gone into some detail on these Jap defenses because many readers of the *Journal* will undoubtedly be working on similar fortifications later on. The Okinawa defense may well be typical of what we can expect on the home islands of Japan."[3] The time and place of the invasion were, however, closely guarded secrets. The first phase of the invasion—Operation Olympic—was tentatively scheduled to take place in southern Kyushu on November 1, 1945.

Like many other Americans, recently sworn-in president Harry Truman worried deeply about the heavy casualties on Okinawa. On June 18, 1945—as the battle drew to a close—Truman met with the Joint Chiefs of Staff, the secretaries of war and the Navy, and others to discuss the next step. The day before the meeting, Truman had written, "I have to decide

Japanese strategy, shall we invade Japan proper or shall we bomb and blockade." At the conclusion of the meeting, Truman stated that its purpose had been to see if "there was a possibility of preventing an Okinawa from one end of Japan to the other." The military chiefs convinced him to proceed with at least the first phase of the invasion scheme.[4] Richard Frank, one of the leading contemporary historians of the final phases of World War II, nevertheless asserted that the Kyushu invasion probably would have been canceled or at least postponed. Radio intelligence from the spring and summer of 1945 was beginning to suggest that the Japanese had identified the probable landing sites and were rapidly rushing reinforcements to the area.[5]

The American planners hoped that the 96th Division would not have to be used for the Kyushu invasion and could instead be saved for the second phase of the invasion of Japan—Operation Coronet. Following the Okinawa campaign, some officers attached to the 96th Division were nonetheless told that the division would be part of the Kyushu invasion in September 1945. Since both that date and the division's stated objective were false, one can reasonably assume that this was part of the general deception plan to keep the Japanese guessing as to the date, place, and units involved in offensive actions following Okinawa.[6] The second landing, scheduled for March 1, 1946, was to be an assault on the main island of Honshu. The goal was to break out of the mountainous beachheads and onto the Kanto Plain near Tokyo, relatively flat terrain where American armor could overwhelm Japanese forces compelled to make a stand in order to defend Tokyo. The 96th Division was scheduled to land five days after the initial assault. As on Okinawa, the 96th would be combined with the 7th and 27th Divisions to form XXIV Corps.[7]

The British also began preparing to fight in Japan. The British fleet expected to participate in the Honshu invasion, and British ground troops expected eventually to participate in the conquest of that island too (even if they were not part of the initial ground landing force). Prime MinisterWinston Churchill projected a loss of 500,000 British lives in the invasion.[8]

In July 1945 Lt. Donald Burrill and some other artillery officers received a briefing on the secret plans for the invasion of Japan. The 96th Division expected to land fifteen to twenty miles from Tokyo.

The briefing officer specified no invasion date. They were told that the amphibious assault would be "tough going." The planned landing zone was in a mountainous region, and the Japanese were fortifying the area with everything they had. The 96th Division would land with the first or second wave. Burrill believed that the Deadeyes would "get their ass shot off."[9]

In addition to Coronet and Olympic, military leaders activated a plan called Pastel, which aimed to deceive the Japanese into thinking that the Americans and British were going to invade China instead of Japan. Part of this plan involved deceiving American troops as well. Historian John Ray Skates noted, for example, that "rumors were intentionally spread among the soldiers of the 6th Army that their next operation would be in China."[10] Lieutenant Walton, although in the Tenth Army, heard this tale as well. Some of the officers in the division were told on a "confidential" basis that the 96th would return to Mindoro to prepare for a landing on the coast of China. After they had landed, they were to fight their way north.[11] Walton sketched the whole "plan" in a 1995 letter. In July 1945, he wrote, "the island [of Okinawa] was secure but cleanup operations were still going on. We were disposing of open ammunition and other item[s] not suitable for our move to Mindoro, P[hilippine] I[slands]. We were told we were going to stage for an invasion of China and move north. Records now show that this was 'misinformation' and that the 96th . . . was going to be in reserve for the main invasion of Japan with an expected total casualties including military and civilians of one million." The actual number of casualties that would have been inflicted during the invasions of Japan is a matter of considerable dispute.[12] The matter can never be satisfactorily resolved because no invasion took place. Few, however, dispute that the casualties would have been high. For the weary veterans of Okinawa, that was a glum prospect indeed.

While land forces prepared for the amphibious assault on Japan, American bombers softened up the Japanese homeland. Lt. Gen. Simon Bolivar Buckner Jr. wrote to his wife just days before his death, "We are now attacking Japan every day from our field here and have already developed our island into a powerful offensive base. When we clean up the Japs here Gen. MacArthur will probably take over our command and I hope he will point us toward Tokyo."[13] Okinawa also served as an emergency landing field for bombers too badly damaged to return to the Marianas.[14]

Redeployment began in July. The 96th Division was bound for its new staging area in the Philippines. In mid-July, refreshed from his first hot showers in months, Lieutenant Walton boarded an LST. The fully loaded convoy remained in Yonabaru harbor for several days to allow a typhoon to pass. The typhoon, however, turned directly toward Okinawa. As the sky darkened, the American convoy made a run for the open sea. Walton vividly recalled the typhoon's impact on his vessel. "It was very rough. The prow of the ship would go up, come down, and smash the water, and then go back up and vibrate up and down in the air. Good thing it was flexible or it could have broken off. One of my friends was on an LST that did break in two, somewhere near Hawaii."[15] As was doubtless the case for many Okinawa veterans, the future military operation weighed heavily on Walton's mind. Even when the weather began to clear, Walton's mood remained somber. "My spirits were not very high as I did not want to go back into combat," he recalled decades later.[16]

On July 16, 1945, the United States successfully tested an atomic bomb in the New Mexico desert. The bomb was not used immediately. In late July 1945 Truman was in Potsdam, Germany, engaged in diplomatic negotiations with Josef Stalin, Winston Churchill, and Clement Attlee. The resultant Potsdam Declaration, issued by the United States, Great Britain, and China on July 26, called for Japan to surrender or face "prompt and utter destruction."[17] Japan, which had already been subjected to firebombing, conventional bombing, and naval blockade, ignored the warning.

On August 6, 1945, the American aircraft *Enola Gay* dropped an atomic bomb on Hiroshima, the headquarters for the Japanese Second Army and the nerve center controlling the defense of Kyushu. An Okinawa veteran who flew low over Hiroshima a few months later described the city, once larger than Fort Worth, Texas, as like a "field in the dead of winter. Absolutely nothing there."[18]

Lieutenant Walton, having been recently promoted to first lieutenant, was at sea when he heard the news of the atomic bomb.

> On the LST going to Mindoro . . . the news was broadcast over the PA system announcing that some new and powerful Atom

> (Atomic) Bomb had been dropped somewhere in Japan. I knew what atoms were, the smallest unit of an element, but how could you make a bomb out of atoms. . . . Regardless of how you make an atomic bomb, I did not want to go back into combat and if the atom bomb . . . could shorten the war it was an answer to prayer for me. The next day we learned the Russians had entered the war against Japan, that also made me happy.[19]

On August 9, 1945, an American bomber dropped a second atomic bomb, on Nagasaki, a port city on the island of Kyushu. Nagasaki was not one of the areas to be seized as part of the upcoming Olympic operation. It was situated on a part of the island scheduled to be sealed off by U.S. forces. In the meantime, it was in the interest of the U.S. forces to neutralize that port. One target at Nagasaki was the Mitsubishi factory where the torpedoes used by the Japanese at Pearl Harbor had reportedly been manufactured.

After the second bomb, the Japanese sued for peace. The recently concluded battle on Okinawa affected the negotiations. Prior to Okinawa, the emperor had sought a decisive victory. Following the Japanese defeat at Okinawa, Hirohito understood that Japan could not continue the war indefinitely. For a time the emperor vacillated by seeking indirect negotiations through the Soviets.[20] In the wake of the atomic bomb and the Soviets' entry into the war, Hirohito's vacillation ended. As a price of their surrender, the Japanese demanded that they be allowed to retain their emperor. Secretary of War Henry Stimson successfully advocated acceptance of the Japanese proposal. Without the existence of a compliant emperor, Stimson argued, Allied forces would have to hunt down and defeat widely scattered Japanese armed forces. It would, he told Truman, lead to "a score of bloody Iwo Jimas and Okinawas all across China and the New Netherlands [today's Indonesia]."[21]

On August 15 Emperor Hirohito announced that Japan would surrender, although he did not use that word. Lieutenant Walton recalled that his outfit was "really jubilant."[22] "That really was an answer to my prayers. I think these events actually took five or six days but we learned of them in three days, probably, after the events had been officially released. We were overjoyed and thought that we would be home in a few months.

Little did I know that it would take eleven more month[s] before my ship sailed under Golden Gate Bridge."[23]

High-level American military leaders already knew about the bomb and the date it was expected to be available when they ordered the Okinawa operation. On August 7, 1944, nearly seven months before the landing on Okinawa, Gen. Leslie Groves, the military director of the atomic bomb project, had promised his superior, Gen. George Marshall, that one type of bomb would be ready by August 1, 1945, and that one or two more could be delivered before December 31, 1945.[24] The Joint Chiefs of Staff, on which General Marshall sat, did not order Admiral Nimitz to seize Okinawa until October 4, 1944, nearly two months after Marshall had learned the projected schedule for the bomb. The directive followed a conference between Adm. Ernest King, the U.S. Navy chief of staff, and Admiral Nimitz during the period September 29–October 1, 1944.[25] Some revisionist historians have suggested that costly battles like Okinawa and Iwo Jima were therefore unnecessary.[26]

Historian Max Hastings calls such interpretations "the knowledge of hindsight."[27] Allied leaders could not have allowed the thousands of military personnel deployed in the Pacific to remain idle while they waited to see if an untried new weapon would decide the outcome of the war. If the project had failed, they would have given the Japanese a respite at the very time the war had turned decidedly against them. Democracies tend to support large-scale conflicts only for limited amounts of time. The seizure of Okinawa allowed the United States to maintain the momentum necessary to bring the war to a close through invasion (in the event that the atomic experiments failed or if the Soviets had proved unwilling to enter the fray against Japan).

Oblivious of future historical debate, the battered Deadeyes basked in the solace of peace. Walton recalled that in August 1945 the division "arrived at Mindoro, PI, unloaded equipment, and established camp on an abandoned airfield [amid] . . . stacks of old planes that had been damaged in combat or in landing, and were not worth repairing."[28] The reprieve from further combat made Mindoro tolerable despite the fact that some veterans considered it a "miserable little island" replete with "jungle rats and mosquitoes."[29] Walton wrote that the men "continued to improve our position using the army philosophy 'move into an area like you will be there a hundred years.' You don't have to do everything the first day, but continually improve your position. We established a place to keep and maintain the 105-mm howitzers in readiness, etc. We laid out a softball field. Set up mess tents with tables. By Christmas we even had a Quonset hut with a steeple [that we used] as a chapel."[30]

On September 2, 1945, the Japanese formally surrendered to the Allied powers. General MacArthur presided over the ceremony on the battleship *Missouri* in Tokyo Bay. The participants in the conflict could now begin to count the cost of the war. The soldiers who were killed or wounded were not the only ones who suffered from the physical impact of combat. When Pfc. Charles Moynihan went back to the battery for medical treatment in mid-June 1945, he was skin and bones. "I was not in very good shape because . . . when I got out of combat I weighed probably 130 or 135 pounds, and I weigh over 200 now and I'm six foot five." Combat took a toll on his spirit too. Moynihan recalled little of the second half of the Okinawa campaign. His mind had simply shut it out.[31]

Lieutenant Walton did not realize the physical impact of combat on his body until he arrived back in the Philippines.

> One day another officer and I had to go into San Jose, the island's capital, to visit some unit which had taken over a sugar mill office for their office. There was a platform scale, so I weighed myself for the first time in a year and found that I weighed 75 kilograms. After some digging for a conversion factor better than 2, I found I weighed 165 pounds, down approximately 15 pounds since senior year in college. "C" rations, powdered eggs, dried potatoes, sleeping on the ground, etc. had kept me pretty trim. Pictures also show that I had the hollow-eyed stare of a recent combat veteran.[32]

Walton was not the only "hollow-eyed" veteran with a blank stare enjoying the respite in comparatively quiet Philippine air.

The Japanese made their greatest artillery effort of the Pacific war on Okinawa. The Americans fought fire with fire. Although historians have largely ignored artillery aspects of the battle in favor of concentrating on infantry and armor, Okinawa constituted the largest artillery battle in America's war against Japan, and artillery played a significant role in the victory. Joseph Alexander noted that "Tenth Army artillery units fired 2,046,930 rounds down range—in addition to 707,500 rockets, mortars, and shells of 5-inch or larger from naval gunfire support," and quoted a Marine lieutenant colonel as saying that the combined Army-Navy-Marine effort "gave us a guns-per-mile-of-front ratio on Okinawa that was higher than any U.S. effort in World War II, similar to the Russian front."[33]

Forward observers played an important role in this massive effort. They served as the central link in the combined arms operation on Okinawa. Mobile American forward observers were a recent development, as were such important innovations as the fire direction control center and the man-portable radio. The battle of Okinawa provides an excellent case study of the impact of these innovations. Since the engagement took place late in the war, the U.S. Army already had resolved many of the problems usually encountered when theory is put into practice. Nonetheless, artillerymen on Okinawa were compelled to deal with new technological innovations—such as the radio-controlled fuse—that both provided new benefits and created new problems.

The mid-level view of the battle of Okinawa presented in this book, based on the oral histories provided by men who were there, differs from most other published accounts, which have focused on high-level views (such as those of General Buckner and Col. Hiromichi Yahara) and very low-level perspectives of the infantry private (E.B. Sledge, Boan, and Dencker). The forward observers were mid-ranking and predominantly middle class, and an in-depth examination of their experiences enriches our understanding of the campaign. Combat conditions imposed new social patterns at variance from routine military hierarchical and courtesy standards. During their three-day rotation to the front lines, officers and enlisted men had a more informal relationship than existed in garrison duty in the United States or even in the rear areas of Okinawa. The officer observers reinforced that informality by removing insignia of rank on the front lines and avoiding actions that might identify them to the Japanese as officers.

As a key component of combined arms operation, these artillerymen needed to interact with other service branches (such as infantry and armor). Within their own branch, they were the only artillerymen who could see what the maneuver commander (infantry or armor) could see. They needed to be cognizant of the needs of low-ranking infantrymen exposed along the front line, the goals of high-ranking officers responsible for conducting the battle of Okinawa, and, most important, the desires of the onsite infantry commander. In general, the observers proved able to interact competently in all those areas.

The Americans' success in maintaining the organizational efficiency of frontline artillery can be explained at least in part by the forward observers' experience with the American civilian economy of the era. "Managerial capitalism"—with its emphasis on middle managers rather than owners—had become an integral part of the U.S. economy in the prewar years.[34] By the time World War II began, the managerial firm—which relied on the decision-making capabilities of middle management—was well on its way to becoming the standard American business organization. Thus the "middle-manager" lieutenants who led the artillery observation teams had grown up in a civilian cultural environment that respected middle-management functions and middle managers. They were well equipped to operate in a team environment in which it was important for each member to feel respected and included.[35] All of the previously mentioned characteristics demonstrated by the observers were consistent with good middle management prevalent in American civilian life in the mid-twentieth century.

During the first half of the twentieth century, Americans gained renown for industrial mass production, with Henry Ford's operation being the prime example. To meet the demands of World War II, these methods were applied not only to the production of war material (such as victory ships), but also to the production of military specialists. One standard "mass production" method was to send a large number of men through training programs while simultaneously recognizing that many of them would eventually fail to qualify for some of the more complex tasks. The U.S. Army's Artillery School at Fort Sill proved capable of mass-producing artillerymen, and that capability contributed to the victory on Okinawa.

The American artillery's ability to outfight the Japanese artillery on Okinawa is due at least in part to Japan's failure to follow a mass-production model. Unlike the American artillery, the Japanese proved unable to rapidly shift artillery fire around the Okinawa battlefield. The roots of that failure can be traced to the Japanese civilian economy. Japan industrialized later than Britain, the United States, and Germany. It was not until after World War II that the Japanese economy was large enough to generate significant modern mass production. Before the war, only a few enterprises based on the managerial capitalism model existed in the entire country.[36] Thus Japan lacked a business culture that fostered middle managers and that could mass-produce large numbers of skilled military specialists. Unlike America, for example, which produced an enormous number of aviators, Japan's military schools were unable to produce a sufficient number of naval aviators to meet wartime demands.[37] Likewise, the Japanese proved unable or unwilling to produce sufficient permanently assigned artillery observers to match the American artillery effort on Okinawa.

The Americans also proved themselves superior to the Japanese in establishing communications between the front line and the artillery support in the rear areas. The Japanese proved incapable of establishing a signals support system adequate for the task of promptly shifting artillery fire about the battlefield. The Americans' superiority in military signals may also be traced to the U.S. civilian economy. During the first half of the twentieth century, two American companies (Western Union and AT&T) operated the world's largest communication networks.[38]

One method of judging the effectiveness of U.S. artillery in the Pacific war is through prisoner-of-war interrogations. Some captured soldiers emphasized the impact of American artillery fire, telling their interrogators that American "artillery shelling and the terrific infantry mortar shelling were having the greatest effect on enemy soldiers of any weapons."[39] The diary of a Japanese soldier found by a member of the 1st Battalion of the 381st reported that American artillery had forced his unit to retreat on May 17, 1945.[40] On May 24 Japanese private Shunji Iwata, captured by the 2nd Battalion on Sugar Loaf Hill, reported that he had been wounded by artillery.[41] Another Japanese army prisoner captured by the 1st Battalion on June 11, 1945, proved to be a Korean impressed into service. "He stated that our Artillery was very effective. also, accounting for many casualties in the rear areas."[42]

Modern high-technology warfare has overshadowed the great infantry battles of the past. The focus now is on longer-distance warfare and scientific developments in weaponry. Many people think that bombing can replace the use of soldiers. Even in the two recent Persian Gulf wars (1991 and 2003), however, air power alone proved incapable of winning the day without the intervention of ground forces. For the foreseeable future, the armed forces will continue to need the services of the infantrymen and the artillery observers who support them. And when the chips are down, as the historian Orlando Davidson said, "it might still be handy to have some troops like the Deadeyes—brave men who will come to grips with the individual enemy and fight it out to the death."[43]

APPENDIX
MEMORIES OF COMBAT LIFE

This appendix contains recollections of the battle of Okinawa that are either too lengthy to quote in the main text or do not fit neatly within the narrative account of the battle. Nonetheless, these vignettes of artillery life on Okinawa are valuable for their historical merit or their humor. Among other things, they illuminate the ambience of the artillery battle. They are presented in roughly chronological order (April–June 1945).

The Stench of War (April 1945)

My schedule was three nights up with the infantry, where we didn't even take our boots off at night and had no spare clothing, and then three nights back at the field artillery battery, where I directed the daylight firing of the 105-mm howitzers. Not taking your clothes off at night, etc., normally was not a problem. However, one day we were advancing through a village and I stepped behind a wall for cover to find that I had stepped into a pigsty. One leg went in above my knee. I had those pants on till I got back to the battery two or three days later. We did have a couple of good rains which washed some of the mess off.

(Lieutenant Walton, Walton correspondence, April 20, 1995)

Hauling the Heavy Artillery Radio

The radio operator had that 610 Radio on his pack saddle on his back. I think that radio weighed seventy pounds. [With] that plus carbine, etc., he must have been carrying about ninety pounds. Another member of my section carried a backpack of four large dry cell batteries that powered the radio when hooked up. The other two members (wire men) were used mainly to change off with the radioman and the battery pack [carrier]. Otherwise they would have been completely exhausted.

On Leyte it was a terrible strain on them. . . . [Forward observer] Lieutenant Fisher (later killed in action) had a water buffalo and had fashioned a saddle to carry the radio. My section had confiscated a dug-out canoe of sorts pulled by a water buffalo. But these two enterprising inventions of the two sections came to an abrupt end. Lieutenant Colonel Mechem (later killed in action) ordered us to abandon our enterprising endeavors to the consternation all involved.

(Lt. Oliver Thompson, Thompson correspondence, June 20, 2009, 3)

Close Calls

Walton had a close and very personal call with Japanese field artillery in the Conical Hill/Yonabaru area: "Shortly before our night march in the mud, I was laying on my cot when our position was suddenly shelled by Japanese field artillery. A shell fragment about a quarter inch in diameter went through the crotch of my pants but did not hit me. Good thing, as if it had been approximately one inch closer, I would not have had children and you [children and grandchildren] would not be reading this letter."

(Lieutenant Walton, Walton correspondence, April 20, 1995)

Lt. Al DeCrans had a close brush with mortar fire. He had moved up front with two soldiers—both very new to forward observation duty. A round from a Japanese knee mortar hit very close to DeCrans and his team. DeCrans felt something hit his chest and saw that both of the inexperienced men had been hit. One was hit in his legs and butt, the other in his side. They got morphine into the two men and sent them to the rear. Then DeCrans looked underneath his shirt and discovered that his binoculars had a big hole in them. That could have been his heart.

(Lieutenant DeCrans, DeCrans interview; DeCrans correspondence, January 29, 2000)

Life in the Mud (May 1945)

In May . . . it started to rain hard on Okinawa. The dirt and gravel roads were not made for the heavy truck and tanks, etc., but they did well while they were dry. Our battery commander [Capt. Rollin Harlow] got the guns moved forward and then the MPs banned all traffic except ambulances on the muddy roads to prevent road destruction. Gun crews had to be provided rations by air drops, as the kitchen and supplies were still in the rear.

Then a day or so later the battery commander rolled in with four M-5 Tractors about 3 or 4 a.m. He said the MPs won't let us on the road during the day but they are not there at night. We have to get the battery together, so load everything up and we will leave about 10 p.m., which meant a night road march with only blackout lights on the vehicles to show your position faintly, but no help to see where you are going. We got everything loaded and all vehicles lashed together in a train fashion: M-5 Tractor (tanklike prime mover), M-10 Trailer, 6'6 GMC [General Motors Corporation] Truck, and then repeated M-5, M-10, GMC.

. . . [W]e started moving in the deep mud with each vehicle pushing and pulling wherever they could at between five and ten miles per hour. Things went well for about an hour, when we came to a big shell hole in the middle of the road. There was a lot of activity with the wounded men being carried on stretchers from ambulances on the frontline side of the shell hole to ambulances on the field hospital side. So we waited. About midnight all activity stopped and the battery commander scouted a route that he thought was passable. So we started very cautiously.

I was in an M-5 behind a GMC that was loaded with kitchen supplies, including the approximately five-foot-high field stove. As we got on the side of the shell hole the truck leaned at such an angle that we all thought it would tip over and possibly take us with it. But it made it and we started to breathe again. So we were on our way again.

One complication was that I had just returned from three days with the infantry and had taken three atabreen [*sic*; malaria prevention] pills to get up to the one-a-day requirement. Unfortunately they have a laxative effect and I had mild diarrhea to the point that once during the trip I had to stick my behind out over the side of the M-5 and let fly. Fortunately no sudden jolts dumped me in the mud below.

We arrived just before daylight, slept about an hour in the vehicles, and then had K rations [crackers and cheese] for breakfast. Then the battery commander told me to get my forward observation party together and he took us up to the infantry front lines.

. . . My forward observation party went forward. The rain continued to fall, to the extent that tanks could not maneuver in the mud. We were on one ridge and the Japanese were on the next ridge in front of us. They had several machine-gun nests along the ridge that we were not able to knock out with mortar or artillery fire. The tank would normally go out and destroy the machine guns just ahead of the infantry, but they could not move so we were stuck.

At night we had our periodic interdictory fire that Captain [Louis] Reuter [Jr., commander of infantry G Company] liked to have placed close to his lines, and naval guns fired parachute flares all night, but somehow the Japanese would sneak up with rifles and grenades for a surprise attack and a pitched battle would rage for about an hour and then suddenly they were gone, resupplied, and would hit the line at a different place.

Captain [Willard G.] Bollinger [commander of infantry F Company] said that they were primarily hand grenade pitching contests and very close-in efforts to overrun our line. One of his men was shot and his body had powder burns on it because of the closeness of the enemy. [My wife] wrote to me and later showed me newspaper clipping describing hand-to-hand and bayonet fighting on Okinawa, but I never saw any.

Due to the mud our jeep could not get up with our replacements, so we had to stay forward for six nights, the longest I have ever gone without taking off my clothes. Using the Army philosophy of "assume you are moving in for a hundred years," we made slow but steady improvements to our five-man slit trench. Normally we slept with four people side by side so close together that you could not turn over. The fifth person would be on guard and sitting up at the end in one-hour time periods. If you wanted to roll over, you needed to do that at the one- or two-hour guard change shuffles.

After digging the trench three or four feet deep we would put up a double-wide poncho tent-type cover using two entrenching shovels as tent poles. This gave an additional foot of headroom and helped the rain run off. We would put a spare poncho on the floor of the hole. At dark we would put on our field jackets, mosquito gloves, and head nets, blow up small (approximately seven inches by ten inches) rubber flotation devices (officially used inside your jacket when wading ashore) for pillows, have our weapons by our side, and go to sleep.

However, with all the rain, the water seeped in and we were sleeping on the water and mud at the bottom of the trench. While I was up on the observation post, some of my men got several 60-mm mortar shell containers (waterproof cardboard cylinders about two inches in diameter and eleven inches long). They carefully placed them in the bottom and laid our poncho over the top. A day or so later a drain was dug so eventually we had a tile-drained slit trench.

The infantry was not as well off. Most of them had open-top trenches with six or more inches of water in them. Many had their rifles suspended on forked sticks about three or four inches above the ground to keep them out of the mud. Those Garand semiautomatic rifles were very good and could be picked up out of the water or mud and still fire.

At night, it sounded like artillery fire was landing very close to our trench, but there was no evidence when daylight came. Then I timed the blast the next night and found that it was some type of echo from my defensive fire that was landing out front. So we were scared by our own fire.

One of the problems was lack of fuel for a very small fire to heat food rations and a canteen cup of water for instant coffee. We quickly burned all the small twigs and sticks. Some people tried to cook with Composition C out of satchel charges, but this powerful plastic explosive burned so quickly and at such a high temperature the food was burned on one side of the C ration can and cold on the other. We eventually burned the asphalt that held the D cells together in our large radio batteries. After those six days my canteen cup had about a one-sixteenth-inch coating of black material that took years to come off.

(Lieutenant Walton, Walton correspondence, May 10, May 18, and May 24, 1995)

The Limited Effectiveness of Air Power

Unfortunately the naval aircraft [on Okinawa] were limited to the five-hundred-pound bombs, which could not destroy the deep (some thirty and forty meters deep) caves and fortifications that the Japanese had built. This was a naval operation under Admiral Nimitz with no Army Air Force planes like the B-17 and B-24 in Europe, which were dropping one-ton blockbuster bombs.

(Lieutenant Walton, Walton correspondence, March 24, 1995)

Binding the Wounds of Battle

On June 26, 1945, Pfc. Fred Goebel wrote Lieutenant "Junior" Walton from his hospital in Guam. Officially, the battle on Okinawa had ended, but the mop-up operations were still in progress. This letter, which serves as the final historical account in this book, demonstrates the human cost of battle to the artillerymen and the familiarity between observation team members despite the difference in rank between the letter writer and the letter recipient.

> Hi ya Junior:
>
> Well here I am once again writing to my old pal. At the present time I'm still laying around in the hospital here at Guam. They let me out of bed about a week ago and am feeling much better now. All I'm waiting from now is a trip to the good old States. As I told you in my previous letter, the nerve in my right arm was severed and will have to go to a nerve center in the States to have it operated on which will probably be in Salt Lake City.
>
> How are the fellows back at the battery getting along lately? In the past week we have had a gathering of Btry "B" here at the hospital. [Pfc. Denver A.] Robinson [wounded June 12, 1945] is in the same ward as I am. [Cpl. Howard E.] Gatenby, who is writing this letter, is in the nut ward because there wasn't enough room for him anywhere else. Also saw Muscles

Brogan and [Pfc. Lonnie] Barnhart. I think Brogan [the battalion mail clerk] is headed back now for the b[at]t[e]ry.

My social environment is limited to shows, reading, cards, chats with my Red Cross girl, who is very pretty by the way, the nurses are out of the question here. Wish you were here so I could borrow a couple of your bars [lieutenant's insignia of rank] and take them out with. That's the only thing the bars are good for anyway. We're going to a lively and exciting game of Bingo tonight. Tomorrow our sewing club gets together oh joy.

We were certainly sorry to hear about Capt. Harlowe [*sic*], and we could hardly believe it when Barnhart [wounded June 16, 1945] told us. It's too bad about [Sgt. Jack] Gilday also but that's war. Well *buddy* I guess this is all for now. Write soon.

(Goebel correspondence, June 26, 1945)

NOTES

Introduction

1. Japan suffered 2.69 million dead during the period 1937–45; China lost 15 to 50 million, the lowest number being the one used here; 5 million died in Japanese-occupied Southeast Asia. The United States lost more than 100,000, and the British had 30,000 dead. Max Hastings, *Retribution: The Battle for Japan, 1944–45* (New York: Alfred A. Knopf, 2008), 541.
2. James H. Hallas, *Killing Ground on Okinawa: The Battle for Sugar Loaf Hill* (Westport, Conn.: Praeger, 1996), xiii; Roy E. Appleman, James M. Burns, Russell A. Gugeler, and John Stevens, *Okinawa: The Last Battle* (Washington, D.C.: Center of Military History, 1948, 1991), 489. The actual number of American deaths during the Okinawa campaign may have been higher than the figure given by Appleman. Donald Dencker, a veteran of the battle and experienced tour guide for the Okinawa battlefields, noted that the Cornerstone of Peace Monument on Okinawa bears the names of 14,005 U.S. dead engraved in stone; see Donald O. Dencker, *Love Company: Infantry Combat against the Japanese: Leyte and Okinawa* (Manhattan, Kans.: Sunflower University Press, 2002), 274.
3. Hastings, *Retribution*, 402, 541; Appleman et al., *Okinawa*, 489; George Feifer, *The Battle of Okinawa: The Blood and the Bomb* (Guilford, Conn.: Lyons Press, 1992, 2001), 408 (originally published as *Tennozan: The Battle of Okinawa and the Atomic Bomb* [New York: Ticknor and Fields, 1992]).
4. David W. Fitz-Simons, "Okinawa: The Last Battle," *Military Review* 76, no. 1 (1996): 77–81, 79.
5. Col. Hiromichi Yahara, *The Battle for Okinawa*, trans. Roger Pineau and Masatoshi Uehara (New York: John Wiley and Sons, 1995), 34.

6. Fitz-Simons, "Okinawa: The Last Battle," 77.
7. Feifer, *The Battle of Okinawa*, 241; Thomas M. Huber, *Japan's Battle of Okinawa, April–June 1945* (Fort Leavenworth, Kans.: Combat Studies Institute, 1990), 74; see also Dencker, *Love Company*, 162, 169, 174, 179, 181.
8. Bob Green, *Okinawa Odyssey: A Texas Rancher's Letters and Recollections of the Battle for Okinawa* (Albany, Tex.: Bright Sky Press, 2004), 119; Hastings, *Retribution*, 377; Capt. Lauren K. Soth, "Cassino of the Pacific: 96th Division Artillery against Strongly-Fortified Okinawa," *Field Artillery Journal* 35 (August 1945): 465–67, 467.
9. Joseph H. Alexander, *Storm Landings: Epic Amphibious Battles in the Central Pacific* (Annapolis: Naval Institute Press, 1997), 170.
10. Hastings, *Retribution*, 380; Gerald Astor, *Operation Iceberg: The Invasion and Conquest of Okinawa in World War II* (New York: Donald I. Fine, 1995), 264, 280–81; Appleman et al., *Okinawa*, 257; Boyd L. Dastrup, *King of Battle: A Branch History of the U.S. Army's Field Artillery* (Fort Monroe, Va.: U.S. Army Training and Doctrine Command, 1992), 225.
11. Huber, *Japan's Battle of Okinawa*, 70, 78.
12. Appleman et al., *Okinawa*.
13. James A. Field Jr., Review of Appleman et al., *Okinawa*, *American Historical Review* 55, no. 2 (1950): 395–96, 395.
14. Chester G. Starr Jr., Review of Appleman et al., *Okinawa*, *Mississippi Valley Historical Review* 36, no. 3 (1949): 548–50, 548.
15. Charles Sheahan interview.
16. Appleman et al., *Okinawa*, 255–66; Green, *Okinawa Odyssey*, 153.
17. One of the more famous books on infantry action is by Marine E. B. Sledge, *With the Old Breed at Peleliu and Okinawa* (Oxford: Oxford University Press, 1981, 1990). See also Dencker, *Love Company*; Jim Boan, *Okinawa: A Memoir of the Sixth Marine Reconnaissance Company* (New York: ibooks, 2000; previously published as *Rising Sun Sinking*); Green, *Okinawa Odyssey*, 153. Some oral accounts such as those by John Garcia and Sledge may be found in Studs Terkel's bestselling *"The Good War": An Oral History of World War Two* (New

York: Ballantine Books, 1984). For a lesser-known account that contains sketches of an infantryman/artist in combat in addition to narrative, see Ken Staley, *The Battle of Okinawa: As Seen from the Foxholes of Ken Staley* (Scotia: Liberty Press, 1994).

18. Starr, Review of Appleman et al., *Okinawa*, 549.
19. Field, Review of Appleman et al., *Okinawa*, 396.
20. Richard L. Watson, Review of Appleman et al., *Okinawa, Journal of Southern History* 15, no. 2 (1949): 275–77, 277.
21. Nicholas Evan Sarantakes, ed., *Seven Stars: The Okinawa Battle Diaries of Simon Bolivar Buckner, Jr., and Joseph Stilwell* (College Station: Texas A&M University Press, 2004), 6–7, 11. While this book focuses primarily on artillery organic to a division, Buckner also made significant use of corps-level artillery. See Janice E. McKenney, *The Organizational History of Field Artillery, 1775–2003* (Washington, D.C.: Center of Military History, 2007), 179–80.
22. Buckner to his wife, April 22, 1945, in Sarantakes, *Seven Stars*, 44.
23. Orlando R. Davidson, J. Carl Willems, and Joseph A. Kahl, *The Deadeyes: The Story of the 96th Infantry Division* (Nashville: Battery Press, 1947, 1981), 195; see also Green, *Okinawa Odyssey*, 117.
24. Sarantakes, *Seven Stars*, 6–7. The words are Sarantakes', not Stilwell's.
25. Davidson et al., *The Deadeyes*, 198; K. P. Jones, *F.O. (Forward Observer)* (New York: Vantage Press, 1989); Williamson Murray and Allan R. Millett, *A War to be Won: Fighting the Second World War* (Cambridge: Belknap Press of Harvard University Press, 2000), 592; Boyd L. Dastrup, *The Field Artillery: History and Sourcebook* (Westport, Conn.: Greenwood Press, 1994), 63.
26. Dastrup, *King of Battle*, 211–12.
27. Soth, "Cassino of the Pacific," 466.
28. See Robert Weiss, *Fire Mission! The Siege at Mortain, Normandy, August 1944* (Shippensburg, Pa.: Burd Street Press, 1998, 2002); James Russell Major, *The Memoirs of an Artillery Forward Observer, 1944–1945* (Manhattan, Kans.: Sunflower University Press, 1999); Eugene Maurey, *Forward Observer* (Chicago: Midwest Books, 1994); Jones, *F.O.*; Edwin V. Westrate, *Forward Observer* (Philadelphia: Blakiston, 1944). For a short article about forward observation during the Guadalcanal campaign of 1942–43, see Capt. John F. Casey,

"An Artillery Forward Observer on Guadalcanal," *Field Artillery Journal* 33 (August 1943): 563–68.

29. A high-level account of corps-level medium artillery published in a technical journal about three months after the battle ended is Col. Bernard S. Waterman, "The Battle of Okinawa: An Artillery Angle," *Field Artillery Journal* 35 (September 1945): 523–28. A typescript titled "Travels of the 362d FA Bn., 96th Inf. Div.: Leyte [and] Okinawa," and apparently written and published by Donald L. McLaughlin, provides rosters for the 362d Field Artillery Battalion.
30. Oliver J. Thompson correspondence, June 16–18, 2009, 5. Lieutenant Thompson served with the 362nd Field Artillery Battalion, a sister battalion of the 361st.
31. The problem of friendly fire on Okinawa is also seldom addressed. Geoffrey Regan's study of military fratricide, *Blue on Blue: A History of Friendly Fire* (New York: Avon Books, 1995), for example, does not mention short-round mishaps during the Okinawa campaign at all. The 96th Division history mentions at least one friendly fire incident near Nishibaru on April 21, 1945; see Davidson et al., *The Deadeyes*, 119.
32. See John Gregor Dunne, "The Hardest War" (review of Sledge, *With the Old Breed at Peleliu and Okinawa*; and Feifer, *The Battle of Okinawa*), *New York Review of Books* 48, no. 20 (December 20, 2001): 50–56, 53.
33. In addition to Sledge's account, other examples focusing on the role of the U.S. Marines include William Manchester, *Goodbye Darkness: A Memoir of the Pacific War* (New York: Dell, 1979, 1980); Hallas, *Killing Ground on Okinawa*; Boan, *Memoir*; Laura Homan Lacey, *Stay off the Skyline: The Sixth Marine Division on Okinawa. An Oral History* (Washington, D.C.: Potomac Books, 2005). Bill Sloan, *The Ultimate Battle: Okinawa 1945—The Last Epic Struggle of World War II* (New York: Simon and Schuster, 2007), provides some Army coverage, but Marine sources predominate. See Thompson correspondence, June 20, 2009, A4, on Army resentment of the focus on Marines.
34. Samuel Eliot Morison, *Victory in the Pacific, 1945* (Boston: Little, Brown, 1960).

35. Perhaps the best Japanese source is the previously cited work of the highest-ranking staff member to survive, operations officer Col. Hiromichi Yahara. Long before he wrote his book (part of which was translated into English), Yahara's view of the battle as given during his interrogation as a prisoner of war influenced U.S. historians such as Appleman and his coauthors. See also Haruko Taya Cook and Theodore F. Cook, *Japan at War: An Oral History* (New York: New Press, 1992), and Huber, *Japan's Battle of Okinawa*. For those who can read Japanese, there is *Memoirs of 15 Years of War* (Naha: Okinawa Prefectural Government, 1996) (written in Japanese except for the testimonies of non-Japanese-speaking witnesses, whose are in English).
36. One forward observer and one battery commander from a different artillery battalion in the 96th Division also provided personal accounts for this study; see Thompson correspondence and John Pfaff interview.
37. Davidson et al., *The Deadeyes*, 195; Dastrup, *King of Battle*, 234; Waterman, "The Battle of Okinawa," 528; Dencker, *Love Company*, 217.
38. For a more technical discussion regarding adjusting artillery fire written in language a layman can understand, see generally Westrate, *Forward Observer*.
39. For leading historical accounts examining the experience of battle in other wars, see, for example, John Keegan, *The Face of Battle* (New York: Barnes and Noble Books, 1976, 1993); and Victor Davis Hanson, *The Western Way of War: Infantry Battle in Classical Greece* (Oxford: Oxford University Press, 1989, 1990). For leading historical accounts of the experience of battle in the European theater of World War II, see, for example, Stephen E. Ambrose, *Citizen Soldiers: The U.S. Army from the Normandy Beaches to the Bulge to the Surrender of Germany, June 1, 1944–May 7, 1945* (New York: Simon and Schuster, 1997); and Stephen E. Ambrose, *D-Day, June 6, 1944: The Climactic Battle of World War II* (New York: Simon and Schuster, 1994, 1995). All of these works focus primarily on the infantry experience.
40. For an expanded discussion of the oral history issues encountered in the preparation of this book, see Rodney Earl Walton, "Memories from the Edge of the Abyss: Evaluating the Oral

Accounts of World War II Veterans," *Oral History Review* 37, no. 1 (2010): 18–34. The section that follows only briefly touches on the issues raised in that article.

41. See Thucydides, *History of the Peloponnesian War*, trans. Rex Warner (New York: Penguin Books, 1954, 1972), 1:22 (p. 48).
42. For an articulate defense of the use of Marine Corps Okinawa veterans' interviews conducted decades after the battle (primarily in 2001–2), see Lacey, *Stay off the Skyline*, 15–19. Although written in layman's language for a popular audience, Lacey's argument is fully documented by numerous authorities. Journalist Bill Sloan conducted his interviews of Okinawa veterans primarily in 2005–6; see Sloan, *The Ultimate Battle*, 379–81.
43. Stephen Everett, *Oral History: Techniques and Procedures* (Washington, D.C.: Center of Military History, 1992), 8, 2.
44. Lacey, *Stay off the Skyline*, 18–19.
45. Ibid., 18.
46. Sheahan correspondence, April 1999 (original in the author's possession).
47. Thompson correspondence, June 16–18, 2009, 7.
48. Davidson et al., *The Deadeyes*. Sheahan praised the accuracy of this account in his interview with me.
49. Battalion History, January 1, 1945, to December 31, 1945, Headquarters 361st Field Artillery Battalion, 96th Infantry Division, National Archives, RG 407, Adjutant General's Office, Unit Operation Reports World War II.
50. For a scholarly discussion of soldiers' ability to recall events early in a campaign (primacy) better than events later in a campaign (recency), see generally Alice M. Hoffman and Howard S. Hoffman, *Archives of Memory: A Soldier Recalls World War II* (Lexington: University Press of Kentucky, 1990), 131, 151; see also Walton, "Memories from the Edge of the Abyss," 27–30.
51. Charles P. Moynihan interview.
52. Alistair Thomson, "Memory as a Battlefield: Personal and Political Investments in the National Military Past," *Oral History Review* 22, no. 2 (1995): 55–73, 61. An abridged version of the article

can be found in Robert Perks and Alistair Thomson, eds., *The Oral History Reader* (New York: Routledge, 1998), 300–310.

53. Everett, *Oral History*, 18.
54. Al DeCrans letter to the author, January 29, 2000.
55. Sloan dedicated his account of the battle to an Okinawa veteran (his late father-in-law) who "never talked about what happened to him there."
56. This opinion is, of course, typical for an oral historian; otherwise they would be using some other method to gather historical information. See John A. Neuenschwander, "Remembrance of Things Past: Oral Historians and Long-Term Memory," *Oral History Review* 6 (1978): 45–53, 47.
57. Although he spent most of his time with the infantry, Moynihan was officially assigned to the Headquarters and Headquarters Battery of the 361st Field Artillery Battalion.
58. For a discussion of the ASTP from the viewpoint of a soldier who participated in the program, see generally Dencker, *Love Company*, especially chapter 2. For an analysis of the program from the viewpoint of a critic who considered the program a luxury and lauded the benefits of the decision to greatly reduce the program, see generally Peter R. Mansoor, *The GI Offensive in Europe: The Triumph of American Infantry Divisions, 1941–1945* (Lawrence: University Press of Kansas, 1999), especially 41–43.
59. Al DeCrans interview. DeCrans recalled this incident taking place north of the port city of Yonabaru, so the date was probably in May 1945. For a similar incident around the same time in which the farm boys of the 382nd Infantry Regiment were pressed into service to butcher a cow on Okinawa, see Dencker, *Love Company*, 247.
60. DeCrans interview.
61. Ibid.
62. Donald Burrill interview; DeCrans interview.
63. Historian William Manchester, a Marine veteran of Okinawa, has been one of the most caustic critics of the 27th Division, which he described as the "infamous Twenty-seventh. They couldn't keep up with the other army units, couldn't even recover their own dead" (Manchester, *Goodbye Darkness,* 411). Lt. Gen. Holland M. Smith,

Marine Corps, advocated disbanding the division; see Holland M. Smith, *Coral and Brass* (New York: Bantam Books, 1948, 1987), 157–58. Boan described the division as being "known as an undisciplined, screw-up outfit" (*Memoir*, 127). Nicholas Sarantakes referred to the 27th as "a substandard National Guard unit"; see Nicholas Evan Sarantakes, "Interservice Relations: The Army and the Marines at the Battle of Okinawa," *Infantry* (January–April 1999): 12–15, 12. The 96th Division personnel interviewed for this project likewise tended to have a negative view of the 27th Division. See, for example, Thompson correspondence, June 20, 2009, A5; Burrill interview; DeCrans interview.

64. Burrill interview.
65. Ibid.
66. For the date of death (March 5, 2007), see *Deadeye Dispatch* (Fall 2007), 2.
67. A wartime photograph of Bollinger is in Davidson et al., *The Deadeyes*, 126.
68. William Filter interview. For a description of Bollinger's "classic gamble" on Hacksaw, see Davidson et al., *The Deadeyes*, 126–29.
69. DeCrans interview.
70. Captain Reuter received an incapacitating head wound during the Okinawa campaign. He died in approximately 1998 and was not interviewed for this work. However, his position and relationship with Bollinger are mentioned here because Reuter was prominently mentioned by three of the narrators. A wartime photograph of Captain Reuter is in Davidson et al., *The Deadeyes*, 126.
71. Willard Bollinger interview.
72. Karel Knutson interview.

Chapter 1. Inventing the American Mobile Artillery Observer

1. Murray and Millett, *War to Be Won*, 592.
2. Shelford Bidwell, *Gunners at War: A Tactical Study of the Royal Artillery in the Twentieth Century* (London: Arms and Armour Press, 1970), 7.
3. Maj. Gen. David E. Ott, "History of the Forward Observer," typescript attached to a letter to Gen. William E. DePuy dated June 25,

1975, pp. A-4-5 in Report (Final Draft), Close Support Study Group, September 12, 1975 (Morris Swett Library, Fort Sill, Okla.). The concept of having forward observers, however, was not new. See Fairfax Downey, *Sound of the Guns: The Story of American Artillery from the Ancient and Honorable Company to the Atom Cannon and Guided Missile* (New York: David McKay, 1955, 1956), 243.

4. Russell Gugeler, "Fort Sill and the Golden Age of Field Artillery," MS, ca. 1981, Morris Swett Library, Fort Sill, Oklahoma, 18. See also McKenney, *The Organizational History of Field Artillery*, 150–55.
5. Brigadier O. F. G. Hogg, *Artillery: Its Origin, Heyday and Decline* (Hamden, Conn.: Archon Books, 1970), 238; J. B. A. Bailey, *Field Artillery and Firepower* (Annapolis: Naval Institute Press, 2004), 141; Boyd L. Dastrup, *King of Battle: A Branch History of the U.S. Army's Field Artillery* (Fort Monroe, Va.: U.S. Army Training and Doctrine Command, 1992), 150.
6. James Russell Major, *The Memoirs of an Artillery Forward Observer, 1944–1945* (Manhattan, Kans.: Sunflower University Press, 1999), xv–xvi.
7. Bailey, *Field Artillery and Firepower*, 141.
8. Dastrup, *King of Battle*, 128.
9. Ibid., 128–29. Guk's book was titled *The Covered Fire of Field Artillery*. Bailey, *Field Artillery and Firepower*, 211; Dastrup, *The Field Artillery*, 180 (includes biographical sketch of Guk).
10. Sir Lawrence Bragg, Maj.-Gen. A. H. Dowson, and Lt.-Col. H. H. Hemming, *Artillery Survey in the First World War* (London: Field Survey Association, 1971), 16; Bidwell, *Gunners at War*, 15–16; Dastrup, *King of Battle*, 129, 149–51.
11. Bailey, *Field Artillery and Firepower*, 272; Dastrup, *King of Battle*, 151; Murray and Millett, *War to Be Won*, 596; John Keegan, *The Face of Battle* (New York: Barnes and Noble Books, 1976, 1993), 264. See also Downey, *Sound of the Guns*, 236, which credits artillery with causing 75 percent of World War I casualties. Bailey, *Field Artillery and Firepower*, 240, 267, credits artillery with causing 58.51 percent of the British casualties and 87 percent of the American casualties. Fred K. Vigman, "The Theoretical Evaluation of Artillery after World War I," *Military Affairs* 16, no. 3 (1952): 115–18, 115.

12. John Keegan, *The First World War* (New York: Vintage Books, 1998, 2000), 22. Keegan commented further on the same situation at 291–92.
13. Bailey, *Field Artillery and Firepower*, 239, 269.
14. Bragg et al., *Artillery Survey*, 11, 16.
15. Bidwell, *Gunners at War*, 15–17.
16. Ibid., 23.
17. Bragg et al., *Artillery Survey*, 11, 16.
18. Robin Prior and Trevor Wilson, *Passchendaele: The Untold Story*, 2nd ed. (New Haven: Yale University Press, 2002), 13.
19. Ibid., 11–12; Bragg et al., *Artillery Survey*, 11, 16; Bailey, *Field Artillery and Firepower*, 270; Edgar F. Raines, *Eyes of Artillery: The Origins of Modern U.S. Army Aviation in World War II* (Washington, D.C.: Center of Military History, 2000), 10–14. Post–World War I analyses, however, declared the aviation observation method "unsatisfactory." See Raines, *Eyes of Artillery*, 15; Downey, *Sound of the Guns*, 216.
20. Gugeler, "Fort Sill," 7.
21. Dastrup, *King of Battle*, 175–76.
22. Keegan, *The First World War*, 374–75. The French had the greater influence on American artillery; see Bailey, *Field Artillery and Firepower*, 266. The United States did not enter the war until 1917, and the bulk of its forces did not arrive in Europe until 1918.
23. Maj. Gen. W. J. Snow in February 1918 as cited in Bailey, *Field Artillery and Firepower*, 266.
24. Dastrup, *King of Battle*, 176. For a further discussion of the limitations on field artillery forward observation in the British army during World War I, see Prior and Wilson, *Passchendaele*, 11–13; Dastrup, *The Field Artillery*, 55, 48, 50.
25. Upton, cited in Bailey, *Field Artillery and Firepower*, 267.
26. Ibid.
27. David T. Zabecki, "Artillery," in *The European Powers in the First World War: An Encyclopedia*, ed. Spencer C. Tucker, 70–76 (New York: Garland, 1996). Bailey (*Field Artillery and Firepower*, 70, 76) noted that the claim by French general Percin in *Le massacre de notre*

infanterie, 1914–1918 (1921) that French artillery had killed 75,000 French soldiers was probably not an experience unique to France.

28. Gugeler, "Fort Sill," 7; Major, *Memoirs*, xvi.
29. Bailey, *Field Artillery and Firepower*, 272.
30. Dastrup, *The Field Artillery*, 48; Major, *Memoirs*, xvi.
31. Bidwell, *Gunners at War*, 227.
32. Gugeler, "Fort Sill," 9–10.
33. Downey, *Sound of the Guns*, 212; Edwin V. Westrate, *Forward Observer* (Philadelphia: Blakiston Company, 1944), 14. The facility at Fort Sill, most often called the Field Artillery School, has had a succession of names and is currently the U.S. Army Fires Center of Excellence and Fort Sill.
34. Gugeler, "Fort Sill," 7.
35. Ott, "History of the Forward Observer," A-4.
36. Gugeler, "Fort Sill," 7–8. For a thumbnail biographical sketch of Brewer, see Dastrup, *The Field Artillery*, 176–77.
37. Gugeler, "Fort Sill," 8, 18; Westrate, *Forward Observer*, 17.
38. Dastrup, *The Field Artillery*, 60, 196; Dastrup, *King of Battle*, 197–99; Gugeler, "Fort Sill," 10–13. For a thumbnail biographical sketch of Ward, see Dastrup, *The Field Artillery*, 191. For a more detailed account of Ward at Fort Sill in the 1930s, see Gugeler, "Fort Sill"; and Capt. Robert O. Kirkland, "Orlando Ward and the Gunnery Department: The Development of the FDC," *Field Artillery* (June 1995): 39–41.
39. Thompson correspondence, June 16–18, 2009, 1–2. For another World War II example of an FDC refusing fire (in this case despite repeated FO requests and continuing American casualties), see Jones, *F.O.*, 50. In that European example, one American field army was unsure of the location of the adjacent American field army.
40. Maneuver units are typically infantry and armor formations that seek to close with the enemy. Other types of units support the maneuver units.
41. Larry H. Roberts, "American Field Artillery 1930–1939" (master's thesis, Oklahoma State University, Stillwater, 1977), 36, citing Riley Sunderland, *The History of the Field Artillery School, 1911–1942* (Fort Sill, Oklahoma: Field Artillery School, 1942), 1:184.

42. Roberts, "American Field Artillery," 34–35.
43. Boyd L. Dastrup, command historian, U.S. Army Fires Center of Excellence and Fort Sill, personal communication, December 12, 2007; Roberts, "American Field Artillery," 35.
44. Major, *Memoirs*, xvi; Dastrup, *The Field Artillery*, 60–61.
45. Ott, "History of the Forward Observer," A-4.
46. Dastrup, *King of Battle,* 203–5; Murray and Millett, *War to Be Won*, 30; Downey, *Sound of the Guns*, 239–40; Lt. Col. Frank G. Ratliff, "The Field Artillery Battalion Fire-Direction Center—Its Past, Present, and Future," *Field Artillery Journal* 40 (May–June 1950): 116–27, 116; and Riley Sunderland, "Massed Fire and the FDC," *Army* (May 1958): 56–59, 56.
47. Sunderland, "Massed Fire," 56. This article was adapted from a manuscript history of the Field Artillery School that Sunderland completed at Fort Sill in 1944.
48. Ott, "History of the Forward Observer," A-4.
49. Downey, *Sound of the Guns*, 243. For a very brief account of the experiences of an Army field artillery forward observation officer on Guadalcanal during December 21, 1942–February 9, 1943, see Capt. John F. Casey, "An Artillery Forward Observer on Guadalcanal," *Field Artillery Journal* 33 (August 1943): 563–68. For the accounts of higher-level Army and Marine artillerymen on Guadalcanal, see Lt. Col. Robert C. Gildart, "Guadalcanal's Artillery," *Field Artillery Journal* 33 (October 1943): 734–39; and Brig. Gen. P. A. del Valle, Marine Field Artillery on Guadalcanal," *Field Artillery Journal* 33 (October 1943): 722–33. For a book-length account of the experiences of an Army field artillery observer in North Africa (using fictitious names and units), see generally Westrate, *Forward Observer*. Brief accounts are also available: Lt. Richard D. Bush, "Forward Observation in Africa," *Field Artillery Journal* 33 (October 1943): 771–75; and Lt. Carl M. Johnstone as told to Maj. Edward A. Raymond, "Up Forward," *Field Artillery Journal* 33 (October 1943): 776–78. For a general overview of artillery developments in the North African campaign, see Dastrup, *King of Battle,* 209–13. Dastrup notes that each field artillery battalion in North Africa had ten or more forward observers to direct fire (p. 211).

50. Ray D. Walton Jr. interview. For the experiences of U.S. Army field artillery observers in Europe, see generally Major, *Memoirs*; Eugene Maurey, *Forward Observer* (Chicago: Midwest Books, 1994); Jones, *F.O.;* and Weiss, *Fire Mission!* (originally published in 1998 as *Enemy North, South, East,*); Major, *Memoirs*, xv, citing Norman Polmar and Thomas B. Allen, eds., *World War II: America at War, 1941–1945* (New York: Random House, 1991), 106.
51. Westrate, *Forward Observer*, 14. Westrate credited Brig. Gen. Jesmond D. Balmer for service as a significant wartime commandant at Fort Sill.
52. Dastrup, *The Field Artillery*, 63.
53. Major, *Memoirs*, xv, citing I. C. B. Dear, ed., *The Oxford Companion to World War II* (Oxford: Oxford University Press, 1995), 57.
54. Weiss, *Fire Mission*, xii–xiii, 213–14.
55. Stephen E. Ambrose, *Citizen Soldiers: The U.S. Army from the Normandy Beaches to the Bulge to the Surrender of Germany, June 1, 1944–May 7, 1945* (New York: Simon and Schuster, 1997), 177.
56. Charles B. MacDonald, *A Time for Trumpets: The Untold Story of the Battle of the Bulge* (New York: Bantam Books, 1984, 1985), 531.
57. Dastrup, *King of Battle*, 236. In 1912, one American artillery officer had written that American artillerymen were so far behind their European counterparts "that it really makes one shudder" (Dastrup, *King of Battle*, 154).

Chapter 2. Prelude to Okinawa

1. Edward S. Miller, *War Plan Orange: The U.S. Strategy to Defeat Japan, 1897–1945* (Annapolis: U.S. Naval Institute, 1991), 21, 24, 151, 156, 161; John A. Lynn, *Battle: A History of Combat and Culture*, rev. ed. (Boulder, Colo.: Westview Press, 2003, 2004), 238, 239.
2. See Murray and Millett, *War to Be Won*, 338, and caption for photo 89; Miller, *War Plan Orange*, 355.
3. Miller, *War Plan Orange*, 330, and the foreword, by Dean C. Allard, xiv.
4. E. B. Potter, *Nimitz* (Annapolis: Naval Institute Press, 1976), 210–11.
5. Miller, *War Plan Orange*, 35, 340 (map 28.1).
6. Eric Bergerud, *Fire in the Sky* (Boulder: Westview Press, 2000), 42.

7. Miller, *War Plan Orange*, 360.
8. Douglas MacArthur, *Reminiscences* (New York: McGraw-Hill, 1964), 145.
9. Miller, *War Plan Orange*, 155, 345, and 340 (map 28.1); see also Murray and Millett, *War to Be Won*, 364; Hastings, *Retribution*, 187.
10. Sheahan interview; Knutson interview.
11. Davidson et al., *The Deadeyes*, 21.
12. Knutson interview; Klimkowicz interview.
13. Thompson correspondence, June 16–18, 2009.
14. Sheahan interview; Sheahan correspondence, April 1999.
15. Sheahan interview; Thompson correspondence, June 16–18, 2009, 4–5.
16. Filter interview.
17. Murray and Millett, *War to Be Won*, 372.
18. Davidson et al., *The Deadeyes*, 69. According to Donald Dencker, the actual number of Deadeyes killed on Leyte was 481 plus 28 attached troops. Dencker correspondence, February 22, 2009.
19. Filter interview; Hastings, *Retribution*, 178–79.
20. Hastings, *Retribution*, 175, 178–79, 187, 191; Murray and Millett, *War to Be Won*, 370, 372.
21. Davidson et al., *The Deadeyes*, 69.
22. Sheahan interview.
23. Hastings, *Retribution*, 249.
24. CINCPOA communiqué 300, March 16, 1945, OPI, *Pacific Fleet Communiques, 1943–1945* (Washington, D.C.: 1945), cited in Murray and Millett, *War to Be Won*, 513.
25. Hastings, *Retribution*, 263; Alexander, *Storm Landings*, 129; Murray and Millett, *War to Be Won*, 513.
26. Miller, *War Plan Orange*, 156–59; Davidson et al., *The Deadeyes*, 79.
27. Dencker, *Love Company*, 146; Bollinger interview.
28. Heath Twitchell, *Northwest Epic: The Building of the Alaska Highway* (New York: St. Martin's Press, 1992), xiv, 46–47, 319.
29. Bollinger interview.
30. Appleman et al., *Okinawa*, 36.
31. APA stands for Attack Personnel Amphibious. Naval writers often refer to this type of vessel simply as an "attack transport"; see, for

example, Samuel Eliot Morison, *Victory in the Pacific, 1945* (Boston: Little, Brown, 1960), xxi.

32. Harold Scott interview. For the description of a similar or identical incident on the convoy headed for Iwo Jima, see James Bradley (with Ron Powers), *Flags of Our Fathers* (New York: Bantam Books, 2000), 136.
33. Walton correspondence, March 17, 1995; *Memoirs of 15 Years of War*, 239.
34. Walton correspondence, March 17, 1995. Although Walton gave the number of this vessel as APA 217, the shipping assignment diagram in Appleman et al., *Okinawa*, chart 4, suggests that it should be APA 117.
35. *Memoirs of 15 Years of War*, 239.
36. Walton correspondence, March 17, 1995.

Chapter 3. Easter Invasion

1. Col. Bernard S. Waterman, "The Battle of Okinawa: An Artillery Angle," *Field Artillery Journal* 35 (September 1945): 523–28, 523.
2. Yahara interrogation report, August 6, 1945, 12–13, in Yahara, *The Battle for Okinawa*, 207–19. Counter-battery fire is artillery directed against enemy artillery positions (rather than against troops, vehicles, or enemy strong points).
3. Waterman, "Battle of Okinawa," 523. Japanese artillery fire fell on American artillerymen on Keise Shima throughout the day of the initial landings on Okinawa proper (April 1, 1945); Sloan, *The Ultimate Battle*, 35–37.
4. Dastrup, *King of Battle*, 233; Waterman, "Battle of Okinawa," 525.
5. Appleman et al., *Okinawa*, 68–69.
6. Walton correspondence, March 24, 1995.
7. Appleman et al., *Okinawa*, 74.
8. Scott correspondence, May 10, 2008.
9. Scott written statement, June 25, 2008.
10. Scott correspondence, May 10, 2008. Boan (*Memoir*, 33) reported that the vessels *Hinsdale* and LST 884 were hit by kamikazes during the feint assault.
11. Scott correspondence, May 10, 2008.

12. See Alexander, *Storm Landings*, 31.
13. Scott interview.
14. Scott did not recall the specific unit to which these ground forces belonged. The classic account of the Okinawa operation reports that Maj. Gen. Thomas E. Watson's 2nd Marine Division conducted the decoy landing. See Appleman, *Okinawa*, 74 and 30 (map 3).
15. Appleman referred these as the "Minatoga beaches" (Appleman et al., *Okinawa*, 74).
16. Scott interview.
17 Alexander, *Storm Landings*, 155. Okinawa veteran Jim Boan likewise concluded that the feint had frozen "the Japanese main body in a defensive position" (*Memoir*, 55).
18. Appleman et al., *Okinawa*, 7, 69, 75, and map 3.
19. Walton correspondence, March 30, 1995.
20. *Memoirs of 15 Years of War*, 239.
21. Sheahan correspondence, April 1999.
22. Sheahan interview.
23. Burrill interview.
24. Bollinger interview.
25. Knutson interview. "Tom White" is a fictitious name.
26. Ibid.
27. Alexander, *Storm Landings*, 158.
28. Knutson interview; Roman Klimkowicz interview. News of this incident, which apparently took place during the early morning hours of April 2, reached as high as Tenth Army command. On April 3, 1945, General Buckner noted in his diary: "Yesterday 11 women armed and in Jap uniforms along with some soldiers tried to attack 96th Div FA [field artillery] positions and were killed" (Sarantakes, *Seven Stars*, 31).
29. Walton correspondence, March 30, 1995.
30. Ibid.
31. Ibid.
32. Ibid. The device was named after the vessel on which it operated—the LST *Brodie*. See Raines, *Eyes of Artillery*, 267 and 269, for a picture of the *Brodie* with the device installed. Raines (p. 270) described the Brodie device as "the major tactical innovation of the Okinawa

campaign." It was on Okinawa that the most significant use of the Brodie device took place. Also see Ken Wakefield (in association with Wesley Kyle), *The Fighting Grasshoppers: U.S. Liaison Aircraft Operations in Europe, 1942–1945* (Leicester: Midland Counties Publications, 1990), 19. Wakefield asserts that an artillery lieutenant named James H. Brodie invented the device.

33. Walton correspondence, March 30, 1995.
34. American intelligence officer Rear Adm. Robert N. Colwell wrote that Cho "had a reputation for being somewhat impatient, even to the point of being hot headed"; see Colwell, "Intelligence and the Okinawa Battle," *Naval War College Review* 38, no. 2 (1985): 81–95, 92.
35. James Belote and William Belote, *Typhoon of Steel: The Battle for Okinawa* (New York: Harper and Row, 1970), 18–19, 327.
36. Ibid., 20; see also Yahara, *Okinawa.*
37. Huber, *Japan's Battle of Okinawa*, 5–6.
38. Thompson correspondence, June 20, 2009, 2.
39. Potter, *Nimitz*, 326–27. The conference took place in San Francisco from September 29 to October 2, 1944.
40. Huber, *Japan's Battle of Okinawa*, 10, 13, 28; Walton videotaped interview, December 29, 1993; Boan, *Memoir*, 94.
41. Huber, *Japan's Battle for Okinawa*, 118.
42. Buckner's leadership ability in general has proved controversial. Retired Marine colonel Joseph Alexander described Buckner as "a popular, competent commander . . . [whose] experience with amphibious warfare had been limited to observing the Aleutian landings" (*Storm Landings*, 165). Retired Marine Corps Reserve colonel Allan R. Millett (and his coauthor) characterized Buckner as "hardly fit to command a corps, let alone a field army" (*War to Be Won*, 515).
43. William Manchester, *American Caesar: Douglas MacArthur, 1880–1964* (New York: Dell, 1978, 1983); 504; D. Clayton James, *The Years of MacArthur: 1941–1945* (Boston: Houghton Mifflin, 1975), 733.
44. Walton correspondence, May 6, 1945. Lieutenant Walton was prohibited by censorship restrictions from discussing the battle in letters home—a restriction that did not seem to apply to General Buckner.

Chapter 4. Assault on Kakazu Ridge

1. Sheahan correspondence, April 1999.
2. Moynihan interview.
3. Robert Leckie, *Okinawa: The Last Battle of World War II* (New York: Penguin Group, 1995), 78.
4. Filter interview.
5. Moynihan interview.
6. Burrill interview.
7. Sheahan interview.
8. See, for example, the fictional exchange between a disgruntled infantry company commander and a new forward observer in North Africa in Westrate, *Forward Observer*, 46.
9. Walton correspondence, April 9, 1995.
10. Walton correspondence, April 9 and 12, 1995.
11. Colwell, "Intelligence and the Okinawa Battle," 93.
12. Sheahan interview and correspondence.
13. Davidson et al., *The Deadeyes*, 103, 106, 108.
14. Ibid., 104.
15. Filter interview.
16. Davidson et al., *The Deadeyes*, 109.
17. Moynihan interview. The "buzz bomb" was probably a Japanese 320-mm spigot mortar, which had a range of roughly three-quarters of a mile. Moynihan may have meant to say "one thousand yards" rather than "one thousand feet." Huber, *Japan's Battle of Okinawa*, 75.
18. Moynihan interview.
19. Ibid.
20. 381st Unit Report no. 7, April 11, 1945, 2.
21. For a detailed description of the events as they unfolded, see Davidson et al., *The Deadeyes*, 109.
22. Ibid.
23. Filter interview.
24. Moynihan interview.
25. Ibid.
26. Sheahan interview; Sheahan correspondence, April 1999.
27. 381st Unit Report no. 7, April 11, 1945, 2.
28. Davidson et al., *The Deadeyes*, 109.

29. Walton correspondence, April 12, 1995. Walton later received a Bronze Star for valor as a result of his conduct on April 10, 1945, and a Purple Heart as a result of his wound.
30. Walton correspondence, April 12, 1995; Hastings, *Retribution*, 379.
31. Moynihan interview.
32. Davidson et al., *The Deadeyes*, 109.
33. For comparisons between the Okinawa campaign and World War I by Americans who fought in the campaign, see Soth, "Cassino of the Pacific," 466; and Sledge, *With the Old Breed at Peleliu and Okinawa*, 147, 215, 261, 265. For secondary sources, see Huber, *Japan's Battle of Okinawa*, 64, and Hastings, *Retribution*, 377–78.
34. Huber, *Japan's Battle of Okinawa*, 64, citing in part Eric J. Leed, *No Man's Land: Combat and Identity in World War I* (London: Cambridge University Press, 1979), 130, 139–40.
35. Bollinger interview.
36. Sheahan interview.
37. Ibid.
38. Davidson et al., *The Deadeyes*, 112. A carbine is a small, light rifle used as a sidearm by soldiers with other responsibilities (mortar men, officers, etc.).
39. Moynihan interview.
40. Ibid.
41. Ibid.
42. Filter interview.
43. Moynihan interview.
44. Davidson et al., *The Deadeyes*, 113.
45. Moynihan interview.
46. Sarantakes, *Seven Stars*, 159, n. 66.
47. Alexander, *Storm Landings*, 164; Boan, *Memoir*, 48, 111, 120, 126. Maj. Jon T. Hoffman likewise blamed kamikazes for the ammunition problems in "The Legacy and Lessons of Okinawa," *Marine Corps Gazette* 79, no. 4 (April 1995): 64–71, 69.
48. Moynihan interview.
49. 381st S-4 log, April 10 and April 11, 1945.
50. Moynihan interview.
51. 381st Unit Report 11, April 15, 1945.

52. Although Roosevelt died on April 12 in the United States, the event occurred during the early morning hours of April 13 on Okinawa, on the other side of the International Date Line. See Morison, *Victory in the Pacific*, 230–31.

Chapter 5. Daily Life

1. Ott, "History of the Forward Observer," A-4.
2. Knutson interview; McKenney, *The Organizational History of Field Artillery*, 159; George Forty, *US Army Handbook, 1939–1945* (New York: Barnes and Noble, 1995), 75, 73.
3. Forty, *Handbook*, 75, 73.
4. Moynihan interview.
5. Dastrup, *King of Battle*, 236.
6. For a narrative account of some of the informal artillery-infantry discussions (with the names changed) in North Africa concerning improving the system, see Westrate, *Forward Observer*, 81–82, 117, 131, and 141. For an account of U.S. Army artillery forward observation on Guadalcanal, see generally Capt. John F. Casey, "An Artillery Forward Observer on Guadalcanal," *Field Artillery Journal* 33 (August 1943): 563–56.
7. This team configuration differed only slightly from the North African experience described in Westrate, *Forward Observer*, 12. See also Jones, *F.O.*, 34, 61. Officially the World War II teams were only authorized three men—one officer (observer), one radioman, and one wireman. McKenney, *The Organizational History of Field Artillery*, 182; Major, *Memoirs*.
8. Jones, *F.O.*, 34–35.
9. Soth, "Cassino of the Pacific," 466. Brig. Gen. Richard Gard's precise comments on this point are quoted in their entirety later in this chapter.
10. Ibid. The U.S. Army suffered from this type of shortage in places other than Okinawa, as McKenney pointed out in *The Organizational History of Field Artillery*, 182, 189.
11. McKenney, *The Organizational History of Field Artillery*, 182.
12. Walton videotaped interview, December 29, 1993. All four of B Battery's observer teams were designated 242 and were differentiated only by the name of the observer.

13. Pfaff interview.
14. Thompson correspondence, August 5, 2009; Walton videotaped interview, December 29, 1993; Knutson interview; Klimkowicz interview.
15. Knutson interview.
16. DeCrans interview.
17. Thompson correspondence, August 5, 2009.
18. Moynihan interview.
19. Captain Sheahan served primarily as the liaison officer and forward observer for the 1st Battalion of the 381st Infantry on Okinawa, although at times his work overlapped with the 2nd and 3rd Battalions (Sheahan correspondence, April 1999). For further discussions of the problems transporting the radio, see the Appendix.
20. Moynihan interview.
21. Sheahan correspondence, April 1999.
22. McKenney, *The Organizational History of Field Artillery*, 182; Moynihan interview.
23. Burrill interview.
24. Sheahan interview.
25. Ibid.
26. Thompson correspondence, June 20, 2009, 11. For the 361st conforming to expectations, see the Moynihan interview.
27. Sheahan interview. Forward observers in Europe reported a similar phenomenon; see Major, *Memoirs*, 74. On the informal nature of the relationship between officers and men on Okinawa, see the June 8, 1945, correspondence exchanged between Lieutenant Walton and Pfc. Fred Goebel quoted in chapter 8, and Goebel's June 26, 1945, letter in the Appendix.
28. DeCrans interview; Major, *Memoirs*, 27–28.
29. DeCrans interview. For a social history of the regular Army in the decades before World War II, see generally Edward M. Coffman, *The Regulars: The American Army, 1898–1941* (Cambridge: Belknap Press of Harvard University Press, 2004).
30. DeCrans interview; DeCrans correspondence, January 29, 2000; Dencker, *Love Company*, 250.
31. Moynihan interview. "Green" is a fictitious name used to avoid any embarrassment that might arise from this minor incident.

32. DeCrans interview; DeCrans correspondence, January 29, 2000. See chapter 7 on the significance of Conical Hill.
33. Hall survived the war. In the late 1990s he was living in Seymour, Texas. Gilday survived the war and was believed to be living in the Philadelphia area in the late 1990s.
34. DeCrans interview; Klimkowicz interview; Davidson et al., *The Deadeyes*, 269.
35. Huber, *Japan's Battle of Okinawa*, 78–79.
36. Sheahan interview.
37. DeCrans interview.
38. Walton correspondence, April 20, 1995; see also DeCrans interview.
39. Jones, *F.O.*, 56; Moynihan interview; Knutson interview; Soth, "Cassino of the Pacific," 467.
40. Klimkowicz interview; Knutson interview.
41. Moynihan interview. "Smith" is a fictitious name.
42. Sheahan correspondence, April 1999.
43. Bollinger interview.
44. Feifer, *The Battle of Okinawa*, 235. For a similar observation, see Soth, "Cassino of the Pacific," 465; Boan, *Memoir*, 127.
45. Bollinger interview. Usually two of the 2nd Battalion's three rifle companies would be in the front line and the other would be a bit behind in reserve. Captain Bollinger generally had to share an observation team with the other infantry company that was on the front line.
46. Walton correspondence, April 20, 1995.
47. Waterman, "The Battle of Okinawa," 523–28, 528.
48. Soth, "Cassino of the Pacific," 467.
49. DeCrans interview; Walton personal communication; see also Soth, "Cassino of the Pacific," 467.
50. Bollinger interview; Boan, *Memoir*, 23.
51. Bollinger interview; Dencker, *Love Company*, 79, 103.
52. Bollinger interview. As sometimes occurs in oral history, the same event may have two or more slightly different versions. Lieutenant Walton, whose team was up at the front with Captain Bollinger's company the night the first sergeant was killed, remembered that an American lieutenant on the front line next to Walton's observation team had heard a noise and said, "Halt" instead of shooting.

The Japanese soldier who had made the noise hit the dirt and threw a hand grenade, killing the first sergeant. Walton videotaped interview, December 29, 1993, and supplemental interview May 1, 1999. The 381st S-2/S-3 log entry for May 31, 1945 reads: "F Co had 1 man KIA [killed in action] and 2 WIA [wounded in action] when civilians came out of cave and threw a hand grenade in F Co Cp. There were 2 civilians. F Co attempted to stop them and make arrest wehn [*sic*] one threw a hand grenade. 1st Sgt was killed." An entry dated a few minutes later reads: "2 civilians last night have been soldiers [*sic*]. Were carrying stick grenades."

53. DeCrans interview.
54. Moynihan interview.
55. *Memoirs of 15 Years of War*, 239.
56. General Buckner to his wife, May 3, 1945, in A. P. Jenkins, ed., "Lt. Gen. Simon Bolivar Buckner: Private Letters Relating to the Battle of Okinawa," *Ryundai Review of Euro-American Studies* 42 (December 1997): 63–113, 92–93.
57. Sheahan interview; Morison, *Victory in the Pacific*, 217.
58. Moynihan interview. Sometimes the radio would be located in his foxhole and sometimes it would be in the command post (often a tomb or pillbox).
59. Moynihan interview; Sheahan interview.

Chapter 6. April Battles

1. Appleman et al., *Okinawa*, 126, 198.
2. Thompson correspondence, June 20, 2009, 6–7, A2–A3.
3. Ibid.
4. Davidson et al., *The Deadeyes*, 117–18.
5. Burrill interview. Burrill was officially assigned to A Battery, 361st Field Artillery Battalion, 96th Division.
6. Burrill believed that the date was April 18, 1945. There is some discrepancy about the date and location of Burrill's exploits, as discussed in the introduction.
7. Burrill interview. Sometimes teams shrank to one or two men.
8. Ibid.

9. Burrill written statement; Burrill interview. "Chief's" full name was Alfred C. Robertson, and he was one-half Sioux (Davidson et al., *The Deadeyes*, 111). For an account of Robinson's earlier exploits on Kakazu Ridge, see *The Deadeyes*, 111–12.
10. Burrill interview; Burrill written statement.
11. Ibid.
12. Burrill interview.
13. Ibid. An alternative would have been the VT (variable time) fuse, which caused the shell to burst in the air.
14. Burrill written statement.
15. Certificate attached to Burrill's Silver Star recommendation and signed by Capt. John E. Byers, commanding officer of Company B, 381st Infantry.
16. Burrill written statement; Burrill interview. The official Army nomenclature for the Cub aircraft was L-4. "L" stood for "liaison."
17. Burrill interview.
18. Soth, "Cassino of the Pacific," 465–67, 466; McKenney, *The Organizational History of Field Artillery*, 182.
19. Burrill interview.
20. Certificate attached to Burrill's Silver Star recommendation and signed by Capt. John E. Byers.
21. Burrill interview.
22. Certificate attached to Burrill's Silver Star recommendation and signed by Capt. John E. Byers.
23. Burrill interview.
24. Potter, *Nimitz*, 375. This incident took place during Nimitz's visit to Okinawa on April 23, 1945. On the other hand, Buckner's diary entry for April 24 notes only that "Adm. Nimitz left this morning, apparently well pleased" (Sarantakes, *Seven Stars*, 45).
25. Appleman et al., *Okinawa*, 245–47, 265; Feifer, *The Battle of Okinawa*, 180; Davidson et al., *The Deadeyes*, 121.
26. Feifer, *Battle of Okinawa*, 180.
27. For a general description of the terrain on Hacksaw, see Davidson et al., *The Deadeyes*, 123.
28. Filter interview.
29. Moynihan interview.

30. Ibid.; Walton interview. The proximity fuses had been first used for antiaircraft defenses around London and then by field artillery in Europe's battle of the Bulge (late 1944–early 1945). See Astor, *Operation Iceberg*, 264; Dastrup, *King of Battle*, 225. Dencker witnessed the use of these "newly available shells" on Okinawa about the same time—April 19, 1945; see *Love Company*, 193.
31. Moynihan interview.
32. Ibid.
33. Thompson correspondence, June 20, 2009, A1–2; Walton taped message, April 11, 1999.
34. Moynihan interview.
35. General Buckner to his wife, April 27, 1945, in Jenkins, "Lt. Gen. Simon Bolivar Buckner," 63–113, 90–91.
36. Filter interview.
37. 381st S-2/S-3 log, April 26, 1945.
38. For a contemporary description of the infantry combat on Hacksaw (Sawtooth) Ridge, see Capt. Lauren K. Soth, "Hacksaw Ridge on Okinawa," *Infantry Journal* (August 1945): 1–4. Soth apparently interviewed both Bollinger and Reuter. See also Davidson et al., *The Deadeyes*, 121–31.
39. Bollinger interview; Walton videotaped interview, December 29, 1993.
40. Orlando Davidson, "The Alley Fighters of the 96th," *Saturday Evening Post*, March 8, 1947.
41. 381st S-2/S-3 log, April 27, 1945, 8:20 a.m.
42. Moynihan interview.
43. Ibid.
44. Bollinger interview; Walton videotaped interview, December 29, 1993; Sloan, *The Ultimate Battle*, 142–43.
45. 381st S-2/S-3 log dated April 28, 1945, entry for 1350, 1:50 p.m.
46. Bollinger interview.
47. 381st Unit Report 24, April 30, 1945, 2. The unit was authorized 2,961 enlisted men but had only 1,662. The regiment was authorized 153 officers but only had 104.
48. Moynihan interview.

49. 381st S-2/S-3 log, April 29, 1945. The 77th Division had already been engaged in the Okinawa campaign. It had seized the small island of Ie Shima (with its important airfields) just off the coast of Okinawa. Renowned war correspondent Ernie Pyle was killed in that operation.
50. Walton videotaped interview, December 29, 1993.
51. Filter interview.
52. Sheahan correspondence, April 1999. For another account stating that the artillery remained in the line to support other units during the period when the 96th Division was in the rest camp, see the Knutson interview.
53. DeCrans interview; Walton videotaped interview, December 29, 1993.
54. 381st Regimental History, 3.
55. Moynihan interview; Filter interview.
56. Fvilter interview. Major Addy was killed June 20, 1945; see Davidson et al., *The Deadeyes*, 295.

Chapter 7. Reducing the Shuri Line

1. Dencker correspondence, February 25, 2009.
2. See, for example, Sledge, *With the Old Breed at Peleliu and Okinawa*; Feifer, *The Battle of Okinawa*; James H. Hallas, *Killing Ground on Okinawa: The Battle for Sugar Loaf Hill* (Westport, Conn.: Praeger, 1996); and Charles S. Nichols Jr. and Henry I. Shaw Jr., *Okinawa: Victory in the Pacific* (1955; reprint, Nashville: Battery Press, 1989).
3. Appleman et al., *Okinawa*, 356.
4. Davidson et al., *The Deadeyes*, 131; 381st S-2/S-3 log entry for May 10, 1945.
5. Staley, *The Battle of Okinawa*, 3.
6. Sheahan interview; Moynihan interview.
7. Appleman et al., *Okinawa*, 356–57; see also Davidson et al., *The Deadeyes*, 140–41.
8. Sheahan interview. The hilly area referred to as the Yonabaru escarpment is also known as the Chinen Peninsula.
9. "G Co[mpany] CO reported seeing 15 Japs in town. E Co[mpany] patrolled today. Placed 81 mm [mortar] and art[iller]y barrage in

town. CO said it was the best art[iller]y barrage he had ever seen. Will investigate results tomorrow" (381st S-2/S-3 Journal, May 12, 1945, 3).

10. Curt Sprecher interview; 361st Field Artillery Battalion History, January 1, 1945, to December 31, 1945, 5. The division history suggests that the deceased forward observer was 2nd Lt. Walter O. Schwienher; see Davidson et al. *The Deadeyes*, 140. The records of the infantry regiment confirm that G Company received small arms fire from Conical Hill around that time (381st Unit Report 28, May 15, 1945, 1–2, and map overlay). Readers who wish to trace this incident on Appleman's map XXXV should understand that the 2nd Battalion of the 381st had been placed under the operational control of the 383rd Regiment. Thus the 381st does not appear on the map.
11. Davidson et al., *The Deadeyes*, 140; 381st Infantry Regiment "Losses in Action and Accidental Losses," 23; discussions with Sprecher.
12. Filter interview; Bollinger interview; Davidson, *The Deadeyes*, 140.
13. 381st Unit Report 30, May 17, 1945, 1.
14. The Regimental log notes that "G Co[mpany] had 3 wounded by own FA [Field Artillery] short rounds. Approx 25 rounds art[iller]y fell during the day in B[attalio]n area" (381st S-2/S-3 Journal, May 16, 1945, 2); Sheahan interview.
15. 381st Unit Report 30, May 21, 1945, 3. The supporting units were described as follows: "361 FA Bn, in direct support was reinf[orced] by 776 Amph[ibious] T[an]k Bn, 363 FA Bn and 532 FA Bn fires."
16. Yahara interrogation report, August 6, 1945, 8, in Yahara, *The Battle for Okinawa*, 207–19.
17. Huber, *Japan's Battle of Okinawa,* 92.
18. The series of hills on which the Japanese were engaged included Conical, Sugar, and Cutaway. For the Japanese reaction, see the Yahara interrogation report, August 6, 1945, 9, in Yahara, *Battle for Okinawa,* 207–19.
19. Sheahan interview. For a contrary opinion about the success of American night attacks on Okinawa, see Maj. Jon T. Hoffman, "The Legacy and Lessons of Okinawa," *Marine Corps Gazette* 79, no. 4 (1995): 64–71.
20. Waterman, "The Battle of Okinawa," 526. The soldier had been a crewmember for a dual- purpose (air and ground) artillery battery.

21. Ibid., 523–28, 526.
22. Moynihan interview. Moynihan's description implied that the American high command originally hoped to encircle the Shuri Line from the rear by means of a westward thrust toward Naha from the Yonabaru area. However, the Americans then realized that the bulk of the Japanese forces had already escaped. Thus the Americans shifted to pursue the retreating Japanese forces concentrating in the far southern section of the island.
23. Moynihan interview.
24. Sheahan interview.
25. 381st Unit Report no. 47, June 3, 1945, 3; 381st S-2/S-3 log, June 4, 1945, 2; 381st Unit Report no. 50, June 6, 1945, 2.
26. Walton correspondence, May 10, 1995. His detailed written description of this incident is in the Appendix.
27. General Buckner to his wife, May 28, 1945, in Sarantakes, *Seven Stars*, 64; Dencker, *Love Company*, 163.
28. Walton correspondence, May 18 and 24, 1995. His detailed description of the misery of mud during wartime conditions is in the Appendix.
29. Astor, *Operation Iceberg*, 404.

Chapter 8. The Battle Ends

1. 381st S-4 log entries, June 8, 9, and 11, 1945.
2. Davidson et al., *The Deadeyes*, 168ff.; Walton videotaped interview, December 29, 1993.
3. Sheahan correspondence, April 1999; Sheahan interview; Moynihan interview. Moynihan could not remember whether the large escarpment in the south to which he was referring was called the Big Apple.
4. Moynihan interview.
5. Sprecher interview; Filter interview.
6. Appleman et al., *Okinawa*, 456; Hastings, *Retribution*, 444; Herbert P. Bix, "Japan's Delayed Surrender: A Reinterpretation," *Diplomatic History* 19, no. 2 (1995): 197–225, 212.
7. Soth, "Cassino of the Pacific," 466; Moynihan interview.
8. Walton videotaped interview, December 29, 1993; 361st Battalion History, 5, 7. The last name is misspelled as "Boebel."

9. Goebel correspondence, June 8, 1945. Goebel did indeed recover the use of his right arm and became a Washington State highway patrolman after the war.
10. Moynihan interview; 361st Battalion History, 5.
11. Sheahan interview.
12. Moynihan interview.
13. DeCrans interview; Klimkowicz interview.
14. DeCrans interview; DeCrans correspondence, January 29, 2000. DeCrans did not recall the name of the ridge. Davidson et al., *The Deadeyes*, 201.
15. DeCrans interview; DeCrans correspondence, January 29, 2000; Davidson et al., *The Deadeyes*, 269.
16. Moynihan interview; Klimkowicz interview.
17. 361st Battalion History.
18. Goebel correspondence, June 26, 1945. For the complete text of the letter, see the Appendix.
19. Thompson correspondence, June 20, 2009, 4–5.
20. Soth, "Cassino of the Pacific," 466.
21. Green, *Okinawa Odyssey*, 63.
22. Masahide Ota, *This Was the Battle of Okinawa* (Naha: Naha Publishing Company, 1981), 30; Feifer, *The Battle of Okinawa*, 110–11. American forces had encountered numerous Japanese civilians during the capture of Saipan in 1944, but the civilian population there was far smaller than on Okinawa. Additionally, Saipan was officially a post–World War I League of Nations mandate even though in practice Japan treated that island as a possession.
23. 381st Infantry Regiment s-2/S-3 Journal, May 25, 1945, p. 4. American planes had dropped leaflets asking civilians to wear white when on the roads.
24. Sheahan interview.
25. Hastings, *Retribution*, 541; Feifer, *Battle of Okinawa*, 404–6, 408; Bollinger interview; Green, *Okinawa* Odyssey, 136.
26. *Memoirs of 15 Years of War*, 240.
27. For Okinawa as the prelude for the anticipated invasion of Japan, see Ota, *Battle of Okinawa*, 20.

28. Yahara, *The Battle for Okinawa*, 81; Sheahan interview. "Maedera" and "Medeera" are the same village. There is no universally agreed-upon method of transliterating Japanese characters into Latin letters. This account uses the spelling "Medeera."
29. 381st S-2/S-3 log entry, June 15, 1945, at 1806 (6:06 p.m.); Belote and Belote, *Typhoon of Steel*, 306.
30. Thompson correspondence, June 20, 2009, 16–18, and August 5, 2009.
31. 381st Unit Report 65, June 20, 1945, p. 3.
32. Filter interview.
33. DeCrans interview.
34. 381st Unit Report, June 20, 1945, 2.
35. Walton videotaped interview, December 29, 1993. For a more detailed discussion of this rivalry, see Sarantakes, "Interservice Relations," 12–15.
36. Thompson correspondence, June 20, 2009, A3.
37. DeCrans interview; see also Knutson interview. The "lands" (rifling inside the tube of the howitzer) would blow out of the artillery tubes along with the projectiles after heavy use.
38. Knutson interview. Knutson kept copies of two types of leaflet.
39. Ibid.
40. Appleman et al., *Okinawa*, map XIX.
41. Astor, *Operation Iceberg*, 427.
42. Ibid. 204–2, 427; Dencker, *Love Company*, 264; Green, *Okinawa Odyssey*, 121.
43. Astor, *Operation Iceberg*, 100; Soth, "Hacksaw Ridge on Okinawa," 1.
44. Astor, *Operation Iceberg*, 407; Walton videotaped interview, December 29, 1993; Sheahan interview.
45. Yahara, *Battle for Okinawa*, 81.
46. Dencker, *Love Company*, 264, citing an interview with Ishihara.
47. Ken Stinson discussion. For a coral fragment as the immediate cause of death, see Appleman et al., *Okinawa*, 461.
48. Huber, *Japan's Battle of Okinawa*, 115.
49. Alexander, *Storm Landings*, 151.
50. Sheahan interview; Huber, *Japan's Battle for Okinawa*, 116; Appleman et al., *Okinawa*, map XLIX.

51. 381st S-/S-3 log, June 22, 1945.
52. Appleman and his coauthors titled their seminal treatise "Okinawa: The Last Battle." Since the Okinawa campaign officially ended in June and Japan did not sue for peace until mid-August, that designation is controversial. Allied forces continued operating in the Philippines and Southeast Asia. The Soviets, who invaded Japanese-held Manchuria, maintained that the conquest of Hutou on August 26, 1945, was the final battle of World War II. See Glantz, *August Storm*, 177 (map 8-3); Hastings, *Retribution*, 531.
53. 381st Unit Report no. 71, July 2, 1945, Sheahan interview; DeCrans interview.
54. Sledge, *With the Old Breed at Peleliu and Okinawa*, 308.
55. One photo in this book depicts Burrill receiving his award from General Stilwell. Burrill interview. Photographs of the July 3, 1945, ceremony are in Davidson et al., *The Deadeyes*, 186–87. Burrill is not mentioned in the picture captions, but photo 2 shows 1st Sgt. Alfred "Chief" Robertson receiving a Silver Star, likely for heroic conduct on Kakazu Ridge; and photo 7 shows Welch's Cub. See also Davidson, *The Deadeyes*, 111–12. The ceremony took place at Chan airfield near Kamizato. Dencker, *Love* Company, 277–79. For Stilwell's diary entry, see Sarantakes, *Seven Stars*, 92.
56. DeCrans interview; Thompson correspondence, June 20, 2009, 18–19.
57. Bollinger interview; 381st Regimental History, 2; Moyhihan interview.

Chapter 9. Special Topics

1. Wakefield, *The Fighting Grasshoppers*, 6. For a history of the decision to add two spotter aircraft to each field artillery battalion, see pp. 11–19. For a more complete discussion of air-based observers, see Raines, *Eyes of Artillery*; and Wakefield, *The Fighting Grasshoppers*.
2. Green, *Okinawa Odyssey*, 99; Alexander, *Storm Landings*, 160.
3. Alexander, *Storm Landings*, 168.
4. Sheahan interview; Bollinger interview; Burrill interview. Welch and Briggs are discussed in Davidson et al., *The Deadeyes*, 202.
5. Yahara interrogation report, August 6, 1945, 13, in Yahara, *The Battle for Okinawa*, 207–19. The Japanese experience thus resembled that

of the German ground forces in Europe. "The Germans, it was said, moved only their eyeballs when a Cub appeared over the front; to otherwise move was to reveal their position and thereby invite disaster" (Wakefield, *Fighting Grasshoppers*, 6).

6. Huber, *Japan's Battle of Okinawa*, 78–79.
7. Yahara, *Battle for Okinawa*, 24–25. Yahara's opinion is particularly significant because his two immediate superiors, General Ushijima and General Cho, both committed suicide and thus never presented their views.
8. General Buckner to his wife, April 26, 1945, in Jenkins, "Lt. Gen. Simon Bolivar Buckner," 88.
9. DeCrans interview; DeCrans correspondence, January 29, 2000.
10. Huber, *Japan's Battle for Okinawa*, 79; Thompson correspondence, June 20, 2009, A1; Dencker correspondence, February 25, 2009.
11. Knutson interview.
12. Klimkowicz interview. He did not recall the place-name for the road junction's location.
13. Soth, "Cassino of the Pacific," 467.
14. Burrill interview; Green, *Okinawa Odyssey*, 135–36. Burrill could not remember the precise location, but it may have been on the west end of Kakazu Ridge.
15. Moynihan interview; Dencker correspondence, February 25, 2009; Knutson interview.
16. Waterman, "The Battle of Okinawa," 525; Hastings, *Retribution*, 53; Thompson correspondence, June 20, 2009, A1.
17. DeCrans interview.
18. Sheahan interview; Filter interview.
19. Major, *Memoirs*, 72.
20. Thompson correspondence, August 5, 2009.
21. Dencker, *Love Company*, 96. Dencker himself was nearly a casualty of a short American mortar round; see *Love Company*, 234–35.
22. Friendly fire remains a problem in contemporary warfare. At the U.S. Defense Department National Training Center, where "shots" can be monitored electronically to determine their source, fratricide rates during training maneuvers in 1990 were "about 14 percent for

indirect fire and 12 to 15 percent for direct fire" but "had dropped to 7 and 10 percent, respectively" by 1992. Friendly fire during the First Gulf War (1990–91) killed twenty-one U.S. soldiers and wounded sixty-five; Donna Miles, "Fighting Friendly Fire," *Soldiers* 47, no. 9 (September 1992): 34–36. This was a substantial percentage of the U.S. losses because the casualties from that war were light.

23. J. B. A. Bailey, *Field Artillery and Firepower* (Annapolis: Naval Institute Press, 2004), 14.
24. Sheahan interview.
25. Bollinger interview. Bollinger did not recall the name of the ridge or the date.
26. Boan, *Memoir*, 161.
27. Sheahan interview.
28. Ibid.
29. Ibid.
30. 381st Unit Report 45, June 1, 1945, 3.
31. Sheahan interview.
32. Ibid. This shell was a 155-mm shell or a naval shell, neither of which would have been fired by the 361st. However, Sheahan had no idea which American unit fired the shell.
33. Green, *Okinawa Odyssey*, 122, 138.
34. Filter interview.
35. Sheahan interview.
36. DeCrans interview.
37. Pfaff interview. Pfaff was battery commander of B Battery, 362nd Field Artillery Battalion, 96th Division on Leyte and Okinawa. The 362nd was a sister unit of the 361st Field Artillery Battalion and supported a different infantry regiment in the 96th Division.
38. Jones, *F.O.*, 47–48.
39. Major, *Memoirs*, 97–98. The friendly shell fragments pierced Major's canteen.
40. Westrate, *Forward Observer*, 46.
41. Burrill interview. Like Burrill, Captain Byers was awarded a Silver Star on Okinawa. Davidson et al., *The Deadeyes*, 265.
42. Major, *Memoirs*, 118.

43. Walton videotaped interview, December 29, 1993. Jones (*F.O.*, 45) reported an incident in late 1944 Europe when U.S. tanks mistakenly shot up their own command post.
44. Del Valle, "Marine Field Artillery on Guadalcanal," 729.
45. Lt. Richard D. Bush, "Forward Observation in Africa," *Field Artillery Journal* 33 (October 1943): 771–75, 774.
46. Dr. Boyd Dastrup, correspondence, October 1, 2007.
47. Dencker, *Love Company*, 111. For infantry gratitude, see the warm comments that 96th Division infantryman Don Dencker had for Lt. Ralph Palm of the 362nd Field Artillery Battalion in *Love Company*, 88, 195, 227, 235. Also see Thompson correspondence, August 5, 2009.
48. 381st Unit Report 24, April 30, 1945, 3.
49. Sheahan interview; Hastings, *Retribution*, 169; Dencker, *Love Company*, 71–72.
50. 381st S-2/S-3 log, May 26, 1945; Boan, *Memoir*, 184; Green, *Okinawa Odyssey*, 193–94.
51. Casey, "Guadalcanal," 566; Del Valle, "Guadalcanal," 728.
52. Lt. Col. Robert C. Gildart, "Guadalcanal's Artillery," *Field Artillery Journal* 33 (October 1943): 734–39, 737.
53. Thompson correspondence, August 5, 2009, 3.
54. Burrill interview.
55. Thompson correspondence, June 20, 2009, A5–A6. Captain Reuter was one company commander who preferred close-in night defensive fire; Walton interview.
56. Dencker, *Love Company*, 195, 38. Dencker was commenting on friendly fire from the air.
57. Sarantakes, *Seven Stars*, 6–7, 11, 156.
58. Ibid., 11; McKenney, *The Organizational History of Field Artillery*, 158, 187.
59. Sarantakes, *Seven Stars*, 6–7, 75. Of course, the unorthodox Stilwell had his own critics. For the widespread astonishment when Stilwell replaced Buckner as commander of the Tenth Army, see Hastings, *Retribution*, 427.
60. Hastings, *Retribution*, 383.
61. Yahara interrogation report, August 6, 1945, 13.
62. Sloan, *The Ultimate Battle*, 143.

63. 381st S-2/S-3 Journal, April 23, 1945, 2; Bollinger interview; Walton interview.
64. Filter interview.
65. Feifer, *The Battle of Okinawa*, 360–61, 369.
66. Moynihan interview; Knutson interview.
67. Sheahan interview.

Chapter 10. Aftermath

1. Fitz-Simmons, "Okinawa: The Last Battle," 77–81, 81.
2. Alexander, *Storm Landings*, 148. The Spruance quote is on p. 170.
3. Soth, "Cassino of the Pacific," 466.
4. Michael Bess, *Choices under Fire: Moral Dimensions of World War II* (New York: Alfred A Knopf, 2006), 220–21. For the diary entry, see Richard B. Frank, *Downfall: The End of the Imperial Japanese Empire* (New York: Penguin Books, 1999, 2001), 132. For Truman's decision, see Frank, *Downfall*, 143, citing the minutes of the June 18, 1945, meeting.
5. Frank, *Downfall*, 199–213, 276–77, 357–58.
6. Green, *Okinawa Odyssey*, 193; Alexander, *Storm Landings*, 170. The code name for both invasions of Japan was Downfall.
7. John Ray Skates, *The Invasion of Japan: Alternative to the Bomb* (Columbia: University of South Carolina Press, 1994), 204, 206.
8. Ibid., 77; Nicholas Evan Sarantakes, "The Royal Air Force on Okinawa: The Diplomacy of a Coalition on the Verge of Victory," *Diplomatic History* 27, no. 4 (2003): 479–502; 381st S-2/S-3 log entry, June 24, 1945 at 1355 (1:55 p.m.). As will be discussed in a later endnote, much controversy has swirled around the invasion casualty projections of Allied leaders. Some historians interpret large estimates like Churchill's as an after-the-fact justification for using the atomic bomb.
9. Burrill interview; Klimkowicz interview.
10. Skates, *Invasion of Japan*, 162–63.
11. Walton videotaped interview, December 29, 1993. A map indicating the approximate location of the fictitious planned "landings" on the coast of northeast China near Shanghai is in Thomas M. Huber,

Pastel: Deception in the Invasion of Japan (Fort Leavenworth, Kans.: Combat Studies Institute, 1988), 28.

12. Walton correspondence, July 19, 1995. For a discussion of the number of casualties that would have been suffered during the invasions of Japan, see Skates, *Invasion of Japan*, 77–81. For an even more extensive but more controversial discussion of the projected casualty figures for an invasion of Japan, see D. M. Giangreco, "Casualty Projections for the U.S. Invasions of Japan, 1945–1946: Planning and Policy Implications," *Journal of Military History* 61 (July 1997): 521–82.
13. General Buckner to his wife, June 14, 1945, in Sarantakes, *Seven Stars*, 80.
14. Walton correspondence, August 2, 1995.
15. Walton correspondence, August 19, 1995. Dencker, who was also in the 96th Division, specified July 28, 1945, as the date for his vessel's sudden departure because of the typhoon warning; see Dencker, *Love Company*, 286–88.
16. Walton correspondence, July 26, 1995.
17. Hastings, *Retribution*, 470.
18. Green, *Okinawa Odyssey*, 201. Lieutenant Green had a good view of Hiroshima because his pilot dropped down to five hundred feet and circled the ruined city for about ten minutes.
19. Walton correspondence, August 2, 1995.
20. Herbert P. Bix, "Japan's Delayed Surrender: A Reinterpretation," *Diplomatic History* 19, no. 2 (1995): 197–225, 223. The emperor, of course, was not the sole foreign policy decision maker in Japan. Hirohito still faced the delicate task of convincing the Japanese government, army, and navy that his decision was correct.
21. Frank, *Downfall*, 342, citing Stimson's diary entry for August 10, 1945.
22. Walton videotaped interview, December 29, 1993. Walton's views are representative of those of many veterans. Dencker's unit was also at sea when they got the news regarding the atomic bombs. Dencker, *Love Company*, 289. Klimkowicz interview. For recent contributions to the debate over the use of the bomb, see, for example, Hastings, *Retribution*; Frank, *Downfall*; Bess, *Choices under Fire*; and Tsuyoshi

Hasegawa, *Racing the Enemy: Stalin, Truman, and the Surrender of Japan* (Cambridge: Belknap Press of Harvard University Press, 2005).

23. Walton correspondence, August 2, 1995.
24. F. G. Gosling, *The Manhattan Project: Making the Atomic Bomb* (Washington, D.C.: U.S. Department of Energy, 2001), 40.
25. Potter, *Nimitz*, 327; Sarantakes, *Seven Stars*, 156–57, n. 8.
26. Alvin D. Coox, Review of Feifer, *Tennozan*, *Journal of American History* 81 (March 1995): 1814; see also, Hastings, *Retribution*, 542, 403.
27. Hastings, *Retribution*, 542.
28. Walton correspondence, August 9, 1995. Dencker, who appears to have left Okinawa at about roughly the same time as Walton, gave August 11, 1945, for his arrival on Mindoro (*Love Company*, 291–92).
29. Thompson correspondence, October 10, 2009, 4.
30. Walton correspondence, August 23, 1995.
31. Moynihan interview. Moynihan's comments about his inability to recall details about the later part of the Okinawa campaign are quoted verbatim in the introduction.
32. Walton correspondence, August 23, 1995.
33. Alexander, *Storm Landings*, 166–67. For the final sentence, Alexander was quoting Lt. Col. Frederick P. Henderson.
34. Alfred D. Chandler, "The Emergence of Managerial Capitalism," *Business History Review* 58 (Winter 1984): 473–503, 478; Alfred D. Chandler Jr., *The Visible Hand: The Managerial Revolution in American Business* (Cambridge: Belknap Press of Harvard University Press, 1997), 77–78, 188, 196. Chandler's Pulitzer Prize–winning book posits the existence of a managerial revolution in the United States beginning in the second half of the nineteenth century and ending in first half of the twentieth.
35. Chandler, *Visible Hand*, 493, 188.
36. Chandler, "Managerial Capitalism," 479, 501–2.
37. Hastings, *Retribution*, 96.
38. Chandler, *Visible Hand*, 79, 203.
39. Davidson et al., *The Deadeyes*, 196.
40. Translation Report Supplement to 381st Unit Report no. 51, June 7, 1945, item 3.

41. 381st Unit Report no. 38, May 25, 1945, Interrogation Report.
42. Prisoner of War Interrogation Report, Supplement to 381st Infantry Regiment Unit Report dated June 11, 1945, p. 2.
43. Davidson, "The Alley Fighters of the 96th."

BIBLIOGRAPHY

Books

Alexander, Joseph H. *Storm Landings: Epic Amphibious Battles in the Central Pacific*. Annapolis: Naval Institute Press, 1997.

Ambrose, Stephen E. *Citizen Soldiers: The U.S. Army from the Normandy Beaches to the Bulge to the Surrender of Germany, June 1, 1944–May 7, 1945*. New York: Simon and Schuster, 1997.

———. *D-Day, June 6, 1944: The Climactic Battle of World War II*. New York: Simon and Schuster, 1994, 1995.

The Annals of America. Vol. 16: *1940–1949*. Chicago: Encyclopaedia Britannica, 1976.

Appleman, Roy E., James M. Burns, Russell A. Gugeler, and John Stevens. *Okinawa: The Last Battle*. Washington, D.C.: Center of Military History, 1948, 1991.

Astor, Gerald. *Operation Iceberg: The Invasion and Conquest of Okinawa in World War II*. New York: Donald I. Fine, 1995.

Bailey, J. B. A. *Field Artillery and Firepower*. Annapolis: Naval Institute Press, 2004.

Baldwin, Hanson W. *Battles Lost and Won: Great Campaigns of World War II*. New York: Harper and Row, 1966.

Ballantine, Duncan S. *U.S. Naval Logistics in the Second World War*. Princeton: Princeton University Press, 1949.

Belote, James, and William Belote. *Typhoon of Steel: The Battle for Okinawa*. New York: Harper and Row, 1970.

Bergerud, Eric. *Fire in the Sky*. Boulder, Colo.: Westview Press, 2000.

Bess, Michael. *Choices under Fire: Moral Dimensions of World War II*. New York: Alfred A. Knopf, 2006.

Bidwell, Shelford. *Gunners at War: A Tactical Study of the Royal Artillery in the Twentieth Century*. London: Arms and Armour Press, 1970.

Boan, Jim. *Okinawa: A Memoir of the Sixth Marine Reconnaissance Company*. New York: ibooks, 2000. Previously published as *Rising Sun Sinking*.

Bradley, James (with Ron Powers). *Flags of Our Fathers*. New York: Bantam Books, 2000.

Bragg, Sir Lawrence, Maj.-Gen. A. H. Dowson, and Lt.-Col. H. H. Hemming. *Artillery Survey in the First World War*. London: Field Survey Association, 1971.

Chandler, Alfred D., Jr. *The Visible Hand: The Managerial Revolution in American Business*. Cambridge: Belknap Press of Harvard University Press, 1997.

Coffman, Edward M. *The Regulars: The American Army, 1898–1941*. Cambridge: Belknap Press of Harvard University Press, 2004.

Cook, Haruko Taya, and Theodore F. Cook. *Japan at War: An Oral History*. New York: New Press, 1992.

Dastrup, Boyd L. *The Field Artillery: History and Sourcebook*. Westport, Conn.: Greenwood Press, 1994.

———. *King of Battle: A Branch History of the U.S. Army's Field Artillery*. Fort Monroe, Va.: U.S. Army Training and Doctrine Command, 1992.

Davidson, Orlando R., J. Carl Willems, and Joseph A. Kahl. *The Deadeyes: The Story of the 96th Infantry Division*. Nashville: Battery Press, 1947, 1981.

DeCrans, Alfonse P. *"Us Depression Kids, We're Survivors": A Lifetime of Memories and Reflections*. Privately published, ca. 1999.

Dencker, Donald O. *Love Company: Infantry Combat against the Japanese: Leyte and Okinawa*. Manhattan, Kans.: Sunflower University Press, 2002.

Dick, Robert C. *Cutthroats: The Adventures of a Sherman Tank Driver in the Pacific*. New York: Ballantine Books, 2006.

Dower, John. *War without Mercy: Race and Power in the Pacific War*. New York: Pantheon Books, 1986.

Downey, Fairfax. *Sound of the Guns: The Story of American Artillery from the Ancient and Honorable Company to the Atom Cannon and Guided Missile*. New York: David McKay, 1955, 1956.

Everett, Stephen. *Oral History: Techniques and Procedures*. Washington, D.C.: Center of Military History, 1992.

Feifer, George. *Tennozan: The Battle of Okinawa and the Atomic Bomb.* New York: Ticknor and Fields, 1992. Republished as *The Battle of Okinawa: The Blood and the Bomb.* Guilford, Conn.: Lyons Press, 2001.

Forty, George. *U.S. Army Handbook, 1939–1945.* New York: Barnes and Noble, 1995.

Frank, Richard B. *Downfall: The End of the Imperial Japanese Empire.* New York: Penguin Books, 1999, 2001.

Frisch, Michael. *A Shared Authority: Essays on the Craft and Meaning of Oral and Public History.* Albany: State University of New York Press, 1990.

Glantz, David M. *August Storm: Soviet Tactical and Operational Combat in Manchuria, 1945.* Fort Leavenworth, Kans.: Combat Studies Institute, 1983.

Gosling, F. G. *The Manhattan Project: Making the Atomic Bomb.* Washington, D.C.: U.S. Department of Energy, 2001.

Green, Bob. *Okinawa Odyssey: A Texas Rancher's Letters and Recollections of the Battle for Okinawa.* Albany, Tex.: Bright Sky Press, 2004.

Hallas, James H. *Killing Ground on Okinawa: The Battle for Sugar Loaf Hill.* Westport, Conn.: Praeger, 1996.

Hanson, Victor Davis. *The Western Way of War: Infantry Battle in Classical Greece.* Oxford: Oxford University Press, 1989, 1990.

Hasegawa, Tsuyoshi. *Racing the Enemy: Stalin, Truman, and the Surrender of Japan.* Cambridge: Belknap Press of Harvard University Press, 2005.

Hastings, Max. *Retribution: The Battle for Japan, 1944–45.* New York: Alfred A. Knopf, 2008.

Herodotus. *The Histories.* Translated by Aubrey De Sélincourt. London: Penguin Books, 1954.

Hogg, Brigadier O. F. G. *Artillery: Its Origin, Heyday and Decline.* Hamden, Conn.: Archon Books, 1970.

Huber, Thomas M. *Japan's Battle of Okinawa, April–June* 1945. Fort Leavenworth, Kans.: Combat Studies Institute, 1990.

———. *Pastel: Deception in the Invasion of Japan.* Fort Leavenworth, Kans.: Combat Studies Institute, 1988.

James, D. Clayton. *The Years of MacArthur: 1941–1945*. Boston: Houghton Mifflin, 1975.
Jessup, John E., and Robert W. Coakley. *A Guide to the Study and Use of Military History*. Washington, D.C.: Center of Military History, 1988.
Jones, K. P. *F.O. (Forward Observer)*. New York: Vantage Press, 1989.
Keegan, John. *The Face of Battle*. New York: Barnes and Noble Books, 1976, 1993.
———. *The First World War*. New York: Vintage Books, 1998, 2000.
———. *The Second World War*. New York: Penguin Books, 1989, 1990.
Kenney, George C. *General Kenney Reports*. New York: Duell, Sloan and Pearce, 1949, 1987, 1997.
Lacey, Laura Homan. *Stay off the Skyline: The Sixth Marine Division on Okinawa. An Oral History*. Washington, D.C.: Potomac Books, 2005.
Leckie, Robert. *Okinawa: The Last Battle of World War II*. New York: Penguin Group, 1995.
Lynn, John A. *Battle: A History of Combat and Culture*. Rev. ed. Boulder, Colo.: Westview Press, 2003, 2004.
MacArthur, Douglas. *Reminiscences*. New York: McGraw-Hill, 1964.
MacDonald, Charles B. *A Time for Trumpets: The Untold Story of the Battle of the Bulge*. New York: Bantam Books, 1984, 1985.
Mailer, Norman. *The Naked and the Dead*. 1948. Reprint. New York: Henry Holt, 1976, 1981.
Major, James Russell. *The Memoirs of an Artillery Forward Observer, 1944–1945*. Manhattan, Kans.: Sunflower University Press, 1999.
Manchester, William. *American Caesar: Douglas MacArthur, 1880–1964*. New York: Dell, 1978, 1983.
———. *Goodbye Darkness: A Memoir of the Pacific War*. New York: Dell, 1979, 1980.
Mansoor, Peter R. *The GI Offensive in Europe: The Triumph of American Infantry Divisions, 1941–1945*. Lawrence: University Press of Kansas, 1999.
Marshall, S. L. A. *Night Drop: The American Airborne Invasion of Normandy*. New York: Berkeley Publishing Group, 1962, 1986.

Maurey, Eugene. *Forward Observer*. Chicago: Midwest Books, 1994.
McKenney, Janice E. *The Organizational History of Field Artillery, 1775–2003*. Washington, D.C.: Center of Military History, 2007.
Memoirs of 15 Years of War. Naha: Okinawa Prefectural Government, 1996. [In Japanese except for the testimonies of non-Okinawan witnesses, whose are in English.]
Miller, Edward S. *War Plan Orange: The U.S. Strategy to Defeat Japan, 1897–1945*. Annapolis: U.S. Naval Institute, 1991.
Morison, Samuel Eliot. *Victory in the Pacific, 1945*. Boston: Little, Brown, 1960.
Murray, Williamson, and Allan R. Millett. *A War to Be Won: Fighting the Second World War*. Cambridge: Belknap Press of Harvard University Press, 2000.
Nichols, Chas. S. Jr., and Henry I. Shaw Jr. *Okinawa: Victory in the Pacific*. 1955. Reprint. Nashville: Battery Press, 1989.
Ota, Masahide. *This Was the Battle of Okinawa*. Naha: Naha Publishing Company, 1981.
Peak, Donald T. *Fire Mission: American Cannoneers Defeating the German Army in World War II*. Manhattan, Kans.: Sunflower University Press, 2001.
Perks, Robert, and Alistair Thomson, eds. *The Oral History Reader*. New York: Routledge, 1998.
Portelli, Alessandro. *The Death of Luigi Trastulli and Other Stories: Form and Meaning in Oral History*. Albany: State University of New York Press, 1991.
Potter, E. B. *Nimitz*. Annapolis: Naval Institute Press, 1976.
Prior, Robin, and Trevor Wilson. *Passchendaele: The Untold Story*. 2nd ed. New Haven: Yale University Press, 2002.
Raines, Edgar F. *Eyes of Artillery: The Origins of Modern U.S. Army Aviation in World War II*. Washington, D.C.: Center of Military History, 2000.
Regan, Geoffrey. *Blue on Blue: A History of Friendly Fire*. New York: Avon Books, 1995.
Santoli, Al. *Leading the Way: How Vietnam Veterans Rebuilt the U.S. Military. An Oral History*. New York: Ballantine Books, 1993.

Sarantakes, Nicholas Evan, ed. *Seven Stars: The Okinawa Battle Diaries of Simon Bolivar Buckner, Jr., and Joseph Stilwell.* College Station: Texas A&M University Press, 2004.

Skates, John Ray. *The Invasion of Japan: Alternative to the Bomb.* Columbia: University of South Carolina Press, 1994.

Sledge, E. B. *With the Old Breed at Peleliu and Okinawa.* Oxford: Oxford University Press, 1981, 1990.

Sloan, Bill. *The Ultimate Battle: Okinawa 1945—The Last Epic Struggle of World War II.* New York: Simon and Schuster, 2007.

Smith, Holland M. *Coral and Brass.* New York: Bantam Books, 1948, 1987.

Staley, Ken. *The Battle of Okinawa: As Seen from the Foxholes of Ken Staley.* Scotia: Liberty Press, 1994.

Terkel, Studs. *"The Good War": An Oral History of World War Two.* New York: Ballantine Books, 1984, 1985.

Thompson, Paul. *The Voice of the Past: Oral History.* Oxford: Oxford University Press, 1978, 1988.

Thucydides. *History of the Peloponnesian War.* Translated by Rex Warner. New York: Penguin Books, 1954, 1972.

Tolstoy, Leo. *War and Peace.* Translated by Ann Dunnigan. 1868. Reprint. New York: Penguin Group, 1969.

Twitchell, Heath. *Northwest Epic: The Building of the Alaska Highway.* New York: St. Martin's Press, 1992.

Wakefield, Ken (in association with Wesley Kyle). *The Fighting Grasshoppers: US Liaison Aircraft Operations in Europe, 1942–1945.* Leicester: Midland Counties Publications, 1990.

Weiss, Robert. *Fire Mission! The Siege at Mortain, Normandy, August 1944.* 2nd rev., updated ed. Shippensburg, Pa.: Burd Street Press, 2002.

Westrate, Edwin V. *Forward Observer.* Philadelphia: Blakiston, 1944.

Wheeler, Keith. *The Fall of Japan.* Alexandria, Va.: Time-Life Books, 1983.

Yahara, Col. Hiromichi. *The Battle for Okinawa.* Translated by Roger Pineau and Masatoshi Uehara. New York: John Wiley and Sons, 1995.

Articles and Selections

Alexander, Joseph. "Okinawa: The Final Beachhead." *Proceedings U.S. Naval Institute* 121, no. 4 (1995): 78–81.

Bix, Herbert P. "Japan's Delayed Surrender: A Reinterpretation." *Diplomatic History* 19, no. 2 (1995): 197–225.

Bush, Lt. Richard D. "Forward Observation in Africa." *Field Artillery Journal* 33 (October 1943): 771–75.

Casey, Capt. John F. "An Artillery Forward Observer on Guadalcanal." *Field Artillery Journal* 33 (August 1943): 563–68.

Chandler, Alfred D., Jr. "The Emergence of Managerial Capitalism." *Business History Review* 58 (Winter 1984): 473–503.

Colwell, Rear Adm. Robert N. "Intelligence and the Okinawa Battle." *Naval War College Review* 38, no. 2 (1985): 81–95.

Coox, Alvin D. Review of Feifer, *Tennozan. Journal of American History* 81 (March 1995): 1814.

Davidson, Orlando. "The Alley Fighters of the 96th." *Saturday Evening Post*, March 8, 1947.

del Valle, Brig. Gen. P. A. "Marine Field Artillery on Guadalcanal." *Field Artillery Journal* 33 (October 1943): 722–33.

Dingman, Roger. Review of Feifer, *Tennozan. American Historical Review* 99, no. 1 (1994): 323–24.

Dunne, John Gregor. "The Hardest War." Review of Sledge, *With the Old Breed*, and Feifer, *Battle of Okinawa. New York Review of Books* 48, no. 20 (December 20, 2001): 50–56.

Falk, Col. Stanley L. "The Assault on Okinawa." *Army* 45, no. 6 (June 1995): 46–51.

Feifer, George. "The Rape of Okinawa." *World Policy Journal* 17, no. 3 (2000): 33–40.

Field, James A., Jr. Review of Appleman et al., *Okinawa: The Last Battle. American Historical Review* 55, no. 2 (January 1950): 395–96.

Fitz-Simons, David W. "Okinawa: The Last Battle." *Military Review* 76, no. 1 (1996): 77–81.

Giangreco, D. M. "Casualty Projections for the U.S. Invasions of Japan, 1945–1946: Planning and Policy Implications." *Journal of Military History* 61 (July 1997): 521–82.

Gildart, Lt. Col. Robert C. "Guadalcanal's Artillery." *Field Artillery Journal* 33 (October 1943): 734–39.

Hoffman, Maj. Jon T. "The Legacy and Lessons of Okinawa." *Marine Corps Gazette* 79, no. 4 (1995): 64–71.

Itkin, Stanley. Review of Leckie, *Okinawa: The Last Battle of World War II*. *Library Journal* 120 (April 1, 1995): 106.

Jenkins, A. P., ed. "Lt. Gen. Simon Bolivar Buckner: Private Letters Relating to the Battle of Okinawa." *Ryundai Review of Euro-American Studies* 42 (December 1997): 63–113.

Kennedy, Elizabeth Lapovsky. "Telling Tales: Oral History and the Construction of Pre-Stonewall Lesbian History." Reprinted in *The Oral History Reader*, ed. Robert Perks and Alistair Thomson, 355–55. New York: Routledge, 1998.

Kirkland, Capt. Robert O. "Orlando Ward and the Gunnery Department: The Development of the FDC. *Field Artillery* (June 1995): 39–41.

Leonard, Charles J. "A Marine Returns." *After the Battle* 43 (1984): 28–53.

Lewis, Clayton W. "Chronicles of War." Review of E. B. Sledge, *With the Old Breed*. *Sewanee Review* 99 (Spring 1991): 296–302.

Miles, Donna. "Fighting Friendly Fire." *Soldiers* 47, no. 9 (1992): 34–36.

Novicki, Richard S. Review of Stephen Ambrose, *Citizen Soldiers*. *Library Journal* 122 (November 15, 1997): 64.

Ratliff, Lt. Col. Frank G. "The Field Artillery Battalion Fire-Direction Center—Its Past, Present, and Future." *Field Artillery Journal* 40 (May–June 1950): 116–27.

Reed, John S. "Okinawa: The Battle, the Bomb, and the Camera." *Prologue: Journal of the National Archives* 37, no. 2 (2005): 18–23.

Roberts, Larry H. "American Field Artillery, 1930–1939." Master's thesis, Oklahoma State University, Stillwater, 1977.

Sarantakes, Nicholas Evan. "Interservice Relations: The Army and the Marines at the Battle of Okinawa." *Infantry* (January–April 1999): 12–15.

———. "The Royal Air Force on Okinawa: The Diplomacy of a Coalition on the Verge of Victory." *Diplomatic History* 27, no. 4 (2003): 479–502.

Shaffer, Robert. "Misleading Analogies and Historical Thinking: The War in Iraq as a Case Study." *Perspectives on History: Newsmagazine of the American Historical Association* 47, no. 1 (2009): 21–23.

Snyder, Louis L. Review of Appleman et al., *Okinawa: The Last Battle. Journal of Modern History* 33, no. 4 (1961): 482.

Soth, Capt. Lauren K. "Cassino of the Pacific: 96th Division Artillery against Strongly-Fortified Okinawa." *Field Artillery Journal* 35 (August 1945): 465–67.

———. "Hacksaw Ridge on Okinawa." *Infantry Journal* (August 1945): 1–4.

Starr, Chester G., Jr. Review of Appleman et al., *Okinawa: The Last Battle of World War II. Mississippi Valley Historical Review* 36, no. 3 (1949): 548–50.

Sunderland, Riley. "Massed Fire and the FDC." *Army* (May 1958): 56–59.

Thomson, Alistair. "Memory as a Battlefield: Personal and Political Investments in the National Military Past." *Oral History Review* 22, no. 2 (1995): 55–73.

Tzeng, Megan. "The Battle of Okinawa, 1945: Final Turning Point in the Pacific." *History Teacher* 34, no. 1 (2000): 95–118.

Vigman, Fred K. "The Theoretical Evaluation of Artillery after World War I." *Military Affairs* 16, no. 3 (1952): 115–18.

Walton, Rodney Earl. "Memories from the Edge of the Abyss: Evaluating the Oral Accounts of World War II Veterans." *Oral History Review* 37, no. 1 (2010): 18–34.

———. "U.S. Army Front Line Artillery Observation during the 1945 Okinawa Campaign: An Oral History Case Study." Ph.D. diss., Florida International University, Miami, 2009.

Waterman, Col. Bernard S. "The Battle of Okinawa: An Artillery Angle." *Field Artillery Journal* 35 (September 1945): 523–28.

Watson, Richard L. Review of Appleman et al., *Okinawa: The Last Battle. Journal of Southern History* 15, no. 2 (1949): 275–77.

Zabecki, David T. "Artillery." In *The European Powers in the First World War: An Encyclopedia*, ed. Spencer T. Tucker, 70–76. New York: Garland, 1996.

Sound Recordings

"Okinawa Symposium." Presented by the National Museum of the Pacific War, Fredericksburg, Texas, September 17 and 18, 2005. MP3 Audio Recording by Rollin' Recording, Boerne, Texas.

National Archives

Unit Operations Reports World War II, Adjutant General's Office, RG 407, NARA.

361 FABN—After Action Report, April–June 1945. Box 14016 at location 7/37/8/1.

381 INFBN [infantry regiment]—History 1945. Box 14027 at location 7/37/8/2.

381 INFBN [infantry regiment]—Unit Report for Okinawa. Box 14029 at location 7/37/8/2.

Unpublished Manuscripts

Gugeler, Russell. "Fort Sill and the Golden Age of Field Artillery." Ca. 1981. Morris Swett Library, Fort Sill, Okla.

McLaughlin, Donald L. "Travels of the 362nd FA Bn, 96th Inf. Div.: Leyte [and] Okinawa." U.S. Army Military History Institute Library, Carlisle, Pa.

Ott, Maj. Gen. David E. "History of the Forward Observer." Attached to letter to Gen. William E. DePuy, June 25, 1975, as pages A-1–4 and A-1–5. Report (final draft), Close Support Study Group, 12 September 1975. Morris Swett Library, Fort Sill, Okla.

Pfaff, John. "My Military Service." February 4, 2009.

Interviews

Bollinger, Willard G. (ca. 1918–2001). Captain, U.S. Army, Infantry. Company Commander, F Company, 2nd Battalion, 381st Infantry Regiment, 96th Infantry Division. Taped interview March 25, 1999.

Burrill, Donald M. (ca. 1920–ca. 2005). Second lieutenant, U.S. Army, Artillery. Forward observer, A Battery, 361st Field Artillery Battalion, 96th Infantry Division. Taped interview.

Cronshey, Bob. Private, U.S. Army, Infantry. Rifleman, E Company, 2nd Battalion, 381st Infantry Regiment, 96th Infantry Division. Conversation and site visit June 1995.

DeCrans, Alfonse P. Second lieutenant, U.S. Army, Field Artillery. Forward observer, B Battery, 361st Field Artillery Battalion, 96th Infantry Division. Taped interview December 28, 1999.

Filter, William. Private to staff sergeant, U.S. Army, Infantry. Rifleman, Squad Leader, Platoon Leader, G Company, 2nd Battalion, 381st Infantry Regiment, 96th Infantry Division. Taped interview May 13, 1999.

Klimkowicz, Roman (b. 1921). T-5 (technical specialist, 5th class), U.S. Army, Artillery. Wireman, telephone operator, and switchboard operator, B Battery, 361st Field Artillery Battalion, 96th Infantry Division. Taped interview August 22, 2009.

Knutson, Karel. Staff sergeant, U.S. Army, Artillery. Cannoneer to chief of section, battery (also served one three-day tour on a forward observer team on Okinawa), 361st Field Artillery Battalion, 96th Infantry Division. Interview October 6, 1999.

Morrical, Stanley. U.S. Army, Artillery. Number 1 cannoneer for number 2 gun, B Battery, 361st Field Artillery Battalion, 96th Infantry Division. Taped interview December 30, 1999.

Moynihan, Charles P. Private, first class, U.S. Army, Artillery. Artillery liaison team radio operator, Headquarters and Headquarters Battery, 361st Field Artillery Battalion, 96th Infantry Division. Taped interview March 31, 1999.

Pfaff, John (b. 1917). Captain, U.S. Army. Battery commander, B Battery, 362nd Field Artillery Battalion, 96th Infantry Division. Interview July 14, 2009.

Scott, Harold. Lieutenant (junior grade), U.S. Navy. Radar and radio repair officer, USS *Hendry*. Correspondence and discussions 2007–11.

Sheahan, Charles (1920–2009). First lieutenant/captain, U.S. Army, Artillery. Artillery liaison officer, Headquarters and Headquarters Battery, 361st Field Artillery Battalion, 96th Infantry Division. Taped interview March 26, 1999.

Sprecher, Curt (d. 2007). Private, U.S. Army, Infantry. Rifleman, G Company, 2nd Battalion, 381st Infantry Regiment, 96th Infantry Division. Discussions and site visit June 1995; discussions May 7, 1999.

Staley, Ken. Private, U.S. Army, Infantry. Rifleman, K Company, 383rd Infantry Regiment, 96th Division. Conversation June 1995.

Stinson, Ken K. Private, U.S. Marine Corps. Browning Automatic Rifle gunner, 8th Marine Regiment, 4th Marine Division. Lecture and conversation June 15, 2009.

Thompson, Oliver J. (b. ca. 1922). First lieutenant, U.S. Army. Forward observer, 362nd Field Artillery Battalion, 96th Infantry Division. Interview by written questions May–October 2009.

Walton, Ray (b. 1921). Second lieutenant, U.S. Army, Field Artillery. Forward observer, B Battery, 361st Field Artillery Battalion, 96th Infantry Division. Videotaped interview December 29, 1993; verbally supplemented March–May, 1999; site visits 1993 and 1995.

Correspondence

Buckner, Simon B., Jr. (lieutenant general, Tenth Army). Personal papers, numerous letters from 1945.

Burrill, Donald M. No date, but received by the author in approximately February 2000. Recommendation for Silver Star and proposed certificate dated May 26, 1945 (prepared by Capt. John E. Byers, commanding officer of B Company, 1st Battalion, 381st Infantry Regiment).

Dastrup, Boyd L. (command historian, U.S. Army Fires Center of Excellence and Fort Sill). Dated December 12, 2007.

DeCrans, Al. Dated January 29, 2000.

Dencker, Donald O. Dated February 22, 2009; February 25, 2009.

Goebel, Fred. Dated June 8, 1945; June 26, 1945.

Morrical, Stanley. Undated (December 1999).

Scott, Harold. Dated May 10, 2008; June 25, 2008.

Sheahan, Charles. Undated (ca. April 1999).

Thompson, Oliver J. Dated May 27, 2009; June 16–18, 2009; June 20, 2009; October 10, 2009.

Walton, Ray D., Jr. Dated May 6, 1945 (first three pages only, other page[s] apparently missing); March 17, 1995; March 24, 1995; March 30, 1995; April 9, 1995; April 12, 1995; April 20, 1995; May 10, 1995; May 18, 1995; May 24, 1995; July 19, 1995; July 26, 1995; August 2, 1995; August 9, 1995; August 23, 1995.

INDEX

ABOUT THE AUTHOR

Rodney Earl Walton is the son of a U.S. Army field artillery observer who fought in the battle of Okinawa. This book has its roots in his PhD dissertation, which examined the role of those observers. Mr. Walton received his PhD from Florida International University (Miami) in 2009 along with an award for outstanding academic achievement. He currently serves as an adjunct history instructor at the same university, where he has periodically taught courses on World War II and other military subjects since 2001. Mr. Walton served in the U.S. Army from 1969 to 1973, reaching the rank of first lieutenant. He was awarded the Bronze Star for service as a military intelligence officer in Vietnam (1972–73). He is a graduate of Cornell Law School and worked for more than twenty years as a civil litigation attorney in South Florida.

The Naval Institute Press is the book-publishing arm of the U.S. Naval Institute, a private, nonprofit, membership society for sea service professionals and others who share an interest in naval and maritime affairs. Established in 1873 at the U.S. Naval Academy in Annapolis, Maryland, where its offices remain today, the Naval Institute has members worldwide.

Members of the Naval Institute support the education programs of the society and receive the influential monthly magazine *Proceedings* or the colorful bimonthly magazine *Naval History* and discounts on fine nautical prints and on ship and aircraft photos. They also have access to the transcripts of the Institute's Oral History Program and get discounted admission to any of the Institute-sponsored seminars offered around the country.

The Naval Institute's book-publishing program, begun in 1898 with basic guides to naval practices, has broadened its scope to include books of more general interest. Now the Naval Institute Press publishes about seventy titles each year, ranging from how-to books on boating and navigation to battle histories, biographies, ship and aircraft guides, and novels. Institute members receive significant discounts on the Press's more than eight hundred books in print.

Full-time students are eligible for special half-price membership rates. Life memberships are also available.

For a free catalog describing Naval Institute Press books currently available, and for further information about joining the U.S. Naval Institute, please write to:

Member Services
U.S. Naval Institute
291 Wood Road
Annapolis, MD 21402-5034
Telephone: (800) 233-8764
Fax: (410) 571-1703
Web address: www.usni.org